METAPHORS FOR CHANGE

PARTNERSHIPS, TOOLS AND CIVIC ACTIONS FOR SUSTAINABILITY

Edited by Penny Allen
with Christophe Bonazzi and David Gee

American **Penny Allen** is a writer and environmental consultant living in Paris. She has worked on environmental issues in the United States and Europe since 1987, and organised the international conferences ECO 1997, ECO 1999 and ECO 2001. Earlier, Allen worked as a public welfare worker, a university instructor, a community organiser in local land-use planning, a feature film-maker, and, in 2001, she published a non-fiction book, *A Geography of Saints* (Boston, MA: Zoland Books).

Christophe Bonazzi is a Frenchman with a doctorate in ıstrial economics from the Paris École des Mines. He was ‹ary General of the Ecobilan Group from 1992 to 1996 and ‹e then has headed a small, innovative electronics firm. He organised the ECO conferences and is active in local politics in the French Green Party.

Englishman **David Gee** has worked for over two decades on occupational and environmental risk reduction with UK Trade Unions, with the Environmental Group and Friends of the Earth, where he was Director. Since 1995, he has worked for the European Environment Agency in Copenhagen, where he is responsible for emerging issues and scientific liaison. He is now working on the European Environment Agency report, *The Precautionary Principle: Late Lessons from Early Warnings, 1896–2000*.

METAPHORS FOR CHANGE

PARTNERSHIPS, TOOLS AND CIVIC ACTION FOR SUSTAINABILITY

Edited by **PENNY ALLEN** with CHRISTOPHE BONAZZI and DAVID GEE

2001

Special thanks to:

Charlotte Cox in Portland, Oregon
Charlotte Islev in Copenhagen
Nicolas Lichtenstein and Philippe Osset in Paris

Published by Greenleaf Publishing Limited
Aizlewood's Mill
Nursery Street
Sheffield S3 8GG
UK

Based on an original manuscript prepared jointly by the
European Environment Agency (EEA) and the Association for Colloquia on the Environment (ACE)

Typeset by Greenleaf Publishing.
Cover design by Lali Abril.
Printed and bound, using acid-free paper from managed forests, by
The Cromwell Press, Trowbridge, Wiltshire, UK.

British Library Cataloguing in Publication Data:
Metaphors for change in business and the environment :
metaphors, partnerships, tools and civic action for
sustainability
1. Social responsibility of business 2. Sustainable
development 3. Industries - Environmental aspects
4. Environmental management 5. Environmental policy
I. Allen, Penny II. Bonazzi, Christophe III. Gee, David
658.4'08

ISBN 1874719373

CONTENTS

INTRODUCTION

Taking the environment into account modifies all aspects of human activity, including the simplest exchange of goods to industrial and commercial practices, from the conception of products and factories to the growth strategies of major industry, to financial and other relationships between customers and suppliers. The environment is becoming a strategic issue and a key component of competitiveness for many businesses and some nations, such as Denmark and Japan, as their environmental activities begin to move from the margins to the mainstream.

However, political leadership is often lagging behind the cutting edge of business or the public. Certain business and public initiatives appear to be outpacing both political decision-making and administrative directives. Yet, if environmentally focused developments by business and the public are not recognised and communicated to politicians, it becomes more difficult for governments to set higher global objectives within a new socioeconomic framework that supports—rather than hinders—'sustainable development': meeting the needs of today without compromising the ability of future generations to meet their needs.

It is therefore important to illustrate, in an accessible way, how some business and community leaders, along with forward-thinking policy-makers the world over, are incorporating environmental concerns into their practices and therefore into the economy. While we have strong, forward-thinking examples from various industry leaders, it is important to encourage lagging political leaders to be equally courageous and to join the efforts of innovators in creating more 'welfare' from less 'use of nature'.[1]

Although certain business examples are beginning to change the global environment, they are not changing the basic rules of economics, trade and growth that underlie the globalisation of the marketplace. Nor are they having any affect whatsoever on the ever-growing phenomenon of consumerism. Yet such problems as climate change and biodiversity losses cannot be solved without a global change towards sustainable development, an issue that can seem too large, too abstract, too scary to deal with. Therefore, useful ways to get from where we are to where we want to be—metaphors for change—can inspire innovations in the way we meet our needs; and,

1 See David Gee, 'Meeting needs, consuming resources', page 30.

once the scale and direction of change is perceived, then getting there requires new partnerships and tools, new civic actions that re-orient the power balance where needed.

These texts illustrate some of the 'change agents' needed to help us achieve sustainable development: metaphors, partnerships, tools and civic actions for change. What we have here are statements by both major and minor players, all of whom are sincere. They are telling us where they think they are and what they think they are doing. It is for all of us to decide if they are making any difference at all.

Part 1

Metaphors for change

Metaphors for change can deliver to the public and to decision-makers new perceptions ('structured knowledge') that help interpret the past and the present, and help us forge the future. The wider the gap between the 'now' and the 'necessary', the stronger the bridging perceptions have to be in order to break through barriers of fear and conservatism.

Environmental questions have been dealt with metaphorically up to now often by catastrophism or manicheism (zero growth, Malthusianism, Deep Ecology, 'man is the enemy', less is more, etc.). These metaphors have had limited impact because they have failed to connect with the mainstream of cultural, political and business ideas.

Some possibly constructive concepts are:

- **'Sustainable development'** (from the Brundtland Report; WCED 1987), which has failed to widely resonate so far, although the expression, along with 'sustainable growth' and 'sustainability' are now being used by communities, enterprises, etc. to develop long-term plans
- The **'polluter pays principle'**, which has launched new thinking and regulations
- The **'precautionary principle'**, which is not well understood, or clearly defined, and which has generally not led to positive environmental action early enough
- **'Life-cycle thinking'**, which is a paradigmatic change in thinking, inviting consideration of processes or products from cradle to grave

- **'Eco-efficiency'** and **'eco-effectiveness'**, which are practical and suggest improvement by reasonable increments, rather than, say, **'Factor 10'**, which urges an improvement in resource–output ratio to a degree than tends to frighten decision-makers
- **'Circular, not linear'**, **'products to services'** and **'dematerialisation'** are related metaphors that look at production (or at meeting people's needs) in ways with great potential for reducing the flow of stuff bought, consumed, stored or thrown away by consumer society

There are of course other useful metaphors on the horizon, some of them included in this book.

SUSTAINABLE DEVELOPMENT

'Sustainable development' is a problematic expression, a developing metaphor at best. Few people agree on what it means. It is a conservative notion, yet it is transversal, cutting across economic, environmental and social concerns, so everyone thinks it through differently. Anyone can take the term and reinvent it to suit their needs.

One economist who looks at development in such a way as to restore the ethical dimension to the discussion of economic problems is Amartya Sen, Master of Trinity College, Cambridge, and winner of the 1998 Nobel Prize in Economic Science. In his book, *Development as Freedom*, he looks at development as 'an integrated process of expansion of substantive freedoms that connect with one another'. He includes in this process of development the freedom from social loss involved in environmentally wasteful or polluting private industry, a radical notion coming from an economist. The fact that Sen was honoured with the Nobel Prize has thrust his expansive definitions of economics into the spotlight.

Many people aren't waiting around for the definition of sustainable development to gel. The International Environmental Agency for Local Governments (ICLEI)[2] is an association of local governments seeking to build local capacity for sustainable development. Their Local Agenda 21 (LA 21) Initiative was launched at the United Nations Earth Summit in Rio in 1992. As of early 2000, more than 1,800 local governments in 64 countries were working with their communities to prepare LA 21 action plans for sustainable development. They are defining the term by planning for it.

Corporations can take the term 'sustainable development' and redefine it in management terms if they so choose. Is the sustainable growth of a corporation in any way integrated into the sustainability of life on Earth? Is the following presentation by the CEO of DuPont an example of sustainable development?

One of the problems in discussing sustainability is the use of new jargon, such as DuPont's use of 'environmental footprint'. In substituting this expression, it is using a more manageable metaphor than the very complex 'sustainable development'. How can we interpret DuPont's (or anyone's) assertions about sustainability?

It is true that DuPont has an ambitious programme to reduce greenhouse gas emissions and waste, including energy waste, and the company has already reduced green-

2 See www.iclei.org/about.htm.

house gas emissions (nitrous oxide, chlorofluorocarbons, hydrofluorocarbons and perfluorocarbons) by 37% from 1991 through 1997, from more than 275 billion lb, carbon dioxide equivalents, in 1991 to less than 180 billion lb in 1997. DuPont has also reduced its global atmospheric toxic emissions by 65% since 1987 (from 70 million lb to less than 30 million lb by 1997) and its global atmospheric carcinogenic emissions by 82% since 1987 (from 10 million lb to less than 2 million lb by 1997).

As reported to the US Environmental Protection Agency's Toxics Release Inventory, there has been a reduction in DuPont's total toxics waste and emissions of 40% since 1991, down to more than 500 million lb by 1997. In terms of global hazardous waste, there has been an 18% decrease since 1990, bringing the volume to about 2,260.1 million lb in 1997. The statistics, even though improving, remain daunting, much like the notion of 'sustainable development'.

DuPont Chairman and CEO, Chad Holliday, presented the following—DuPont's official take on the prickly sustainability question—at a Greening of Industry Conference held in late 1999 at the University of North Carolina in Chapel Hill. DuPont will be 200 years old in 2002, and today it describes itself as a science company with thousands of products and services based on the fundamental building blocks of chemistry, physics, information science and molecular biology.

SUSTAINABLE GROWTH
DuPont's goal for the 21st century

Chad Holliday

As we think about the next century, we believe our central focus must be on 'sustainable growth'. By this I mean we must create both shareholder and societal value while we reduce our environmental footprint. **Sustainable growth is our operational definition of sustainable development.** We believe growth is a very necessary element of both successful corporations and successful societies, but that growth in the future must be much different than it has been in the past.

Environmental footprint

The greening of industry is related to our concept of environmental footprint. Our definition of environmental footprint is much broader than the traditional industrial measures of wastes and emissions. It includes: injuries and illnesses to our employees and contractors; incidents such as fires, explosions, accidental releases to the environment and transportation incidents; global waste and emissions; and the use of depletable raw materials and energy.

Using footprint as one metric for the greening of DuPont, over the past decade we have maintained our position as one of the safest industrial companies in the world by almost an order of magnitude and we have reduced major incidents from a level of over 100 in the early 1990s to almost zero in 1998 and 1999. We have reduced air toxics by over 60% and air carcinogens by over 80% on a global basis. And we are on track to reduce the greenhouse gas emissions from our global operations (on a carbon-equivalent basis) by 45% by the year 2000, using 1990 as a base year.

The story on waste is also significant. From a classical environmental viewpoint, DuPont has reduced hazardous waste from our global operations by 20% during the 1990s, while production has increased by 35%. However, from a pure business viewpoint, the real story is the increase in first-pass yields (more of the product we want the first time through the system and less waste that needs to be reworked) and uptime of our processes (increases in the capacity of our equipment and much more efficient use of raw materials and energy). In many cases, these have moved from the mid-70% range in the late 1980s to above 90% today. In environmental terms, this is pollution prevention or source reduction. It is one of the reasons we believe that many projects and programmes can be both 'good for business and good for the environment'.

Recently, we have begun to work on the component of environmental footprint that relates to the use of depletable forms of raw materials and energy. For us, this means the use of fossil fuels. In September, we set two major goals for the year 2010 in this area. The first is to source 10% of our global energy needs from renewable energy. The second is to derive 25% of our revenues from non-depletable resources. Both of these goals have significant stretch, but signal our intent to be a much different company in the future.

For most conferences on the greening of industry, this is where the discussion ends. From our standpoint, this is where the discussion should begin. Good environmental stewardship should be a given. But, to achieve sustainable growth, we need to do much more.

Societal value

An emerging area discussed at conferences and in lengthy reports is 'corporate social responsibility'—a very important area with many dimensions that range from charitable giving, to community outreach, to concern for human rights. At DuPont, we view all of these as a part of our societal value but also believe that the primary reason civil society allows us to operate and grow is because of the value we bring through our products and services.

Over the years, our products have been important in the exploration of space, the improvement of agricultural productivity, the protection of lives, the saving of energy, the comfort and appearance of clothing, and the overall quality of life through thousands of different products. More recently, as we have built a broad platform of knowledge in molecular biology, we are seeing the potential to improve the nutrition and health benefits of food, to produce polymers from renewable feedstocks, and to significantly reduce the environmental impact of a broad range of technologies that are commercial today. From this standpoint, we feel good about creating substantial societal value and having increased our societal value over time.

We have also begun to focus on how we can expand the values delivered through our products and services to a broader percentage of the global population. Today, our products and services can reach about one billion of the six billion people on Earth. From a societal viewpoint, our value is limited because our offerings do not meet the real needs of five-sixths of the world population. From a commercial viewpoint, we are not reaching a huge base of potential future customers.

Recently [1999], we have run a series of advertisements on TV and in major newspapers around the world to launch our new corporate positioning, the

'Miracles of Science'. We had found that 'Better Things for Better Living' was no longer adequate to represent DuPont and that we needed a new way to describe what we do.

As part of this campaign, we provide a 'to do' list for the planet which includes many needs—such as generating fresh water from salt water, and growing food in areas of the world where soil conditions are very poor. For some of these we check 'done that', but in others we do not have a way to meet the need but we are working on it. This signals our intent to play a broader role in meeting important needs for all of the world's population. One challenge I will issue to the broad NGO [non-governmental organisation] community is to help us identify and deliver value to the developing economies in a way that provides both societal and shareholder value. We would welcome alliances and partnerships, particularly with organisations that recognise that both values are important to sustainable growth.

Let me give you an example of one such successful partnership. Several years ago we worked with the Carter Center to help fight Guinea Worm disease—a very painful and potential deadly disease that affected over three million people in Africa. Our contribution was the donation of nylon fibre to create simple water filters. Using these filters, and the Carter Center's on-the-ground expertise, the Guinea Worm disease now affects fewer than 70,000 people, and complete eradication is in sight. This example provided substantial societal value. It evoked tremendous positive emotional value from our employees. And it was the right thing to do. And, while we will look for other areas where we can achieve similar results, we must continue to look for ways to do this that also build strong and competitive businesses for the future.

Shareholder value

Let me now turn to shareholder value, a subject not usually discussed in environmental conferences, or only discussed as an adjunct to the environmental message. An important element in the sustainable growth equation for corporations is shareholder value. Any for-profit, publicly owned company in America must concern itself with creating shareholder value or wealth. If it ignores this, the company's owners will eventually revoke its right to operate, just like society will revoke a company's right to operate if its products or processes are deemed harmful. Creating shareholder value therefore goes hand in hand with creating societal value. DuPont is described as a 'blue-chip' company, one that has provided an attractive return to its shareholders over many years. We are one of the original companies on the Dow Jones Industrial index and, today, are the only so-called 'chemical company' still there. Over the past ten years, we have outperformed the S&P 500 Index and provided a cumulative shareholder return of over 400%. And, while we are not yet DuPont.com, we are an attractive investment to people investing for the long term.

As we work on improving both shareholder value and societal value while reducing environmental footprint, we have found a useful metric to help guide our thinking and decisions. This metric is 'shareholder value added per pound of production' or 'SVA/lb'. We developed this metric with the help of several outside people including Stuart Hart, one of the organisers of this conference and a leader in the field of integrating business and the environment. SVA is the shareholder value created above the cost of capital, which typically is 10%–12% for corporations here in the US. On a simplistic basis, shareholder value can be created with both material and with knowledge. The higher the SVA/lb, the greater is the use of 'knowledge intensity' and the lower is the use of 'material intensity' to create economic value. Coupled with the more traditional financial measures such as 'return on invested capital' and 'cash flow', the SVA/lb metric provides a useful and additive guide for portfolio management, and one that is an indicator of the longer-term sustainability of different growth strategies.

At DuPont, we have evaluated the SVA/lb for all of our 80 business units, with some of our businesses beginning to set stretch goals to increase SVA/lb over the next five to ten years. For DuPont as a total corporation, we believe that a stretch, but reasonable, target is a four-fold increase in SVA/lb over the next decade.

Integration is the key

While each of the values—environmental, societal and shareholder—is important in its own right, the real objective of sustainable growth is to develop business strategies that integrate improvement in all three. Let me illustrate our journey to do this using two DuPont businesses.

Our Photopolymer and Electronic Materials (P&EM) business has been serving the electronics and printing industries for the last 30 years. The business uses a common technology platform of polymers that can be manipulated by light. These polymers can create highly patterned printing proofs and printing plates, circuit lines, electronic components and other electronic building blocks to create high-value products for our customers.

Today, P&EM is a $1.3 billion global business with a track record of strong growth in both revenues and earnings. Since 1995, its SVA has grown by over 50% and its SVA/lb is one of the highest of any of our businesses. Its strategic direction as a business is to achieve high-value growth while 'building a smaller footprint'. To do this, the business works with its customers to simplify the entire value chain to reduce process time, energy and material usage by collapsing process steps using our materials. In essence, we create 'smart materials' that greatly simplify the delivery of value.

For example, in the printed wiring board process, we invented the dry film photopolymer resist more than 25 years ago. This product transformed a sloppy wet process using liquid resists and multiple steps to a single dry lamination with orders-of-magnitude better yields.

Today, one of our new products enables customers to digitally image photoresists that are used to make circuit boards. This eliminates the use of many phototools, with all of their associated waste-streams, and allows customers to make circuit boards smaller and smarter, thus using less material and energy.

Our Tyvek® business is a major part of DuPont's Nonwovens Strategic Business Unit which has revenues of about $1 billion. Tyvek® is known for developing strong and durable mailing and courier envelopes used by Federal Express and the US postal service. These envelopes are half the weight of conventional envelopes so they provide energy savings and reduced mailing costs. In addition, Tyvek® envelopes conserve natural resources by using 25% post-consumer recycled content from used milk and water jugs, and the envelopes can be recycled in facilities across the US.

A newer application of Tyvek® is in the construction industry. Tyvek® HomeWrap™ and StuccoWrap™ provide important water and weather protection while homes are under construction and, more importantly, significantly improve the energy efficiency of homes after they are completed. On an annual basis, the homes wrapped with Tyvek® save over ten times the amount of energy used in the total Tyvek® manufacturing operation—a ten-to-one return on the energy investment in the first year. In subsequent years, the homes wrapped with Tyvek® will continue to deliver significant energy savings without any new energy investment.

Finally, the Tyvek® global manufacturing operations have reduced their own environmental footprint by cutting unit energy consumption by 50% over the past five years.

These very brief examples from two of our businesses show that environmental, societal and shareholder values can be integrated into the very fabric of our products and services in a way that creates sustainable, profitable growth. Our strategic intent is to build the DuPont of the 21st century as a series of sustainable growth businesses tied tightly to the core values that have defined our sustainability over the past two centuries.

In closing

We see continuing progress in moving from a focus on the 'greening of industry' to a focus on 'sustainable growth'. This will increasingly require the integration of environmental, societal and shareholder values into all business strategies. The metrics to gauge progress will broaden to include: renewable feedstocks and energy; the societal value of products and services; and creative approaches to differentiating the ways in which shareholder value is created.

DuPont is committed to being a global leader in this critical transformation. However, broad alliances with NGOs can accelerate the process and significantly improve our results. We ask for not only your help with this transformation but that you hold us accountable.

LIFE-CYCLE THINKING
A new metaphor and a new paradigm

Life-cycle thinking has led to a significant turn-of-the-century shift in thinking, both in the way products or processes are perceived as well as in the sharing of responsibility for their creation and use. Author of the following paper, Hélène Teulon, earned her PhD in Industrial Organisation from the École des Mines in Paris. Her speciality is the environmental analysis of automobile design, and she developed a life-cycle inventory for the USCAR consortium—which includes Chrysler, Ford and General Motors—in the first attempt by US car manufacturers to examine the environmental burdens associated with the total life-cycle of a car. She undertook a similar life-cycle inventory of a hypothetical generic car for JAMA, which comprises the Japanese auto-makers. She was Technical Director of the Paris office of Ecobilan when she wrote the following article. She joined the Sustainability Department at PricewaterhouseCoopers in 2000.

LIFE-CYCLE THINKING
What is it?

Hélène Teulon

Our industrial society used to be a collection of independent stakeholders, linked by commercial connections: the brick producer sold bricks to the house builder, who sold houses to the real estate agent, who sold them to customers. Each stakeholder dealt with the problems arising in his or her business. No one was bothered by what happened to the product once delivered. Being able to deliver the product in the requested quantity was the challenge. The growth strategy of companies in this period, characterised as 'fordism', was to standardise products and processes and to increase production capacities.

As production techniques improved, under competitive pressure, each stakeholder started to look further downstream in the production chain, and to integrate the constraints of customers into the design process: the house builder wanted an insulated brick—the brick producer would research the matter and develop a new product; the car manufacturer wanted to reduce its storage capacity—the tyre producer would deliver tyres to the assembly line, just-in-time. With the development of quality procedures and the extension of partnerships within a production chain, this trend spread to a point where a producer would consider not only the requests of direct customers, but those of all the stakeholders involved in the life-cycle of the product: industrials, distributors, users, consumers, and even waste management or recycling companies. This is 'life-cycle thinking'.

When designing or manufacturing a product, a company makes decisions that impact the activities of all the other stakeholders downstream in the chain: cost, quality, technical performance and environmental impacts. Life-cycle thinking consists of considering the different phases of the product life-cycle before making any decisions related to the product, service or process under study.

Life-cycle thinking: an efficient approach to business issues

Life-cycle thinking is based on a well-known mathematical theorem: the global optimum is not the sum of the local ones. If each stakeholder improves his or her own business performance, without considering the whole chain, then the product might in the end be less competitive than it could have been if each operation had been planned in relationship to the others. This makes sense for costs, technical performance or environmental performance. As such, life-cycle thinking is the efficient way of doing business in the long term.

For example, an industrialist can use a cheaper thermoset plastic instead of technical thermoplastic for a particular part and discover later that this choice is problematical with regard to recycling when the piece is eventually discarded after use, thermoplastic being more easily recycled than thermoset plastic. Such examples are abundant. Car manufacturers must decide, in choosing materials for the car body, between steel and aluminium. Aluminium is lighter and thus leads to reduced fuel consumption during the car's life-cycle, but it consumes much more energy during its manufacture, at least if it is primary aluminium. Choices must be made on a case-by-case basis, considering the particular part, the life-span of both the part and the car, and other technical and economic considerations.

Another basic principle of life-cycle thinking is that a change in the product use or end-of-life is usually hundreds of times cheaper if it is planned ahead of time at an upstream stage of the life-cycle. For example, the European Directive draft on end-of-life of vehicles states that parts containing hexavalent chromium must be dismantled before any end-of-life treatment occurs. It is probably much cheaper to find alternative design options for cars than it would be to dismantle the currently chromed parts. And it is probably cheaper to do this at an early development stage than at the final development step of a new car.

Life-cycle thinking as a framework

Life-cycle thinking has provided a framework for a family of tools that have been developed during the last 15 years. For example:

- Life-cycle assessment (LCA), as the name implies, involves accounting for the flows of all materials and energy entering and exiting the system defined by the life-cycle of a product or a service (life-cycle inventory), then calculating potential impacts of these flows on the environment, and then interpreting the results in order to improve, at the lowest cost, the environmental performance of the product throughout its life-cycle (ISO 1997).
- Life-cycle costing involves a method of accountability that looks at the whole chain rather than at a single part of the chain.
- Design for environment, design for recycling and design for dismantling involve providing designers and inventors of products the tools and information they need to integrate environmental constraints, including

dismantling and recycling, into the development of the product (Aggeri and Hatchuel 1998).

- Eco-innovation involves a qualitative method to encourage the innovation necessary for solving environmental questions, essentially through brainstorming sessions focused on a life-cycle thinking approach (Fussler with James 1996).

Figure 1 shows how these tools lead to the implementation of eco-conception. In most cases, the following elements are present:

- The assessment of environmental regulations
- The systematic and quantified exploration of each phase of the life-cycle on all aspects of the environment (LCA)
- Pragmatic methods such as design for disassembly and design for recycling

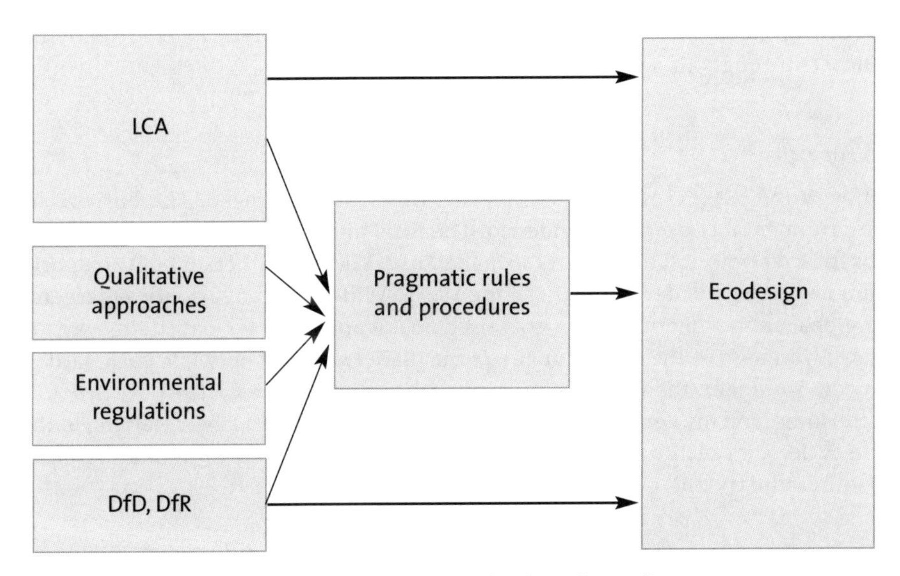

LCA = life-cycle assessment; DfD = design for dismantling; DfR = design for recycling

Figure 1 **Different tools in the life-cycle thinking framework**

Life-cycle thinking is under way

Although all companies are not yet involved in life-cycle thinking, there is no doubt that this trend is growing in many different industrial sectors. Some companies clearly state that they promote life-cycle thinking. This is the case for

Philips Electronic Products, for example, which launched a life-cycle thinking programme a few years ago. This is also the case for DaimlerChrysler, where a 'life-cycle management' programme was established in the mid-1990s. Other companies, if they do not call what they are doing life-cycle thinking, nevertheless use tools and methods that are directly related to the life-cycle thinking framework, such as life-cycle assessment, life-cycle costing and design for environment.

At the same time, the International Organisation for Standardisation (ISO) is working to standardise life-cycle assessment, in its 14040 standards series. Similarly, the organisation, in its work on 'Type III' environmental labelling, clearly states that 'the full life-cycle of the product or service' should be considered (ISO 1998).

Further, regulating authorities are now adopting the notion of life-cycle thinking. European eco-labels explicitly mention the life-cycle of the product and cite life-cycle assessment as an outstanding tool for measuring the impacts of products and services on the environment. The European Commission is preparing the Integrated Product Policy which will depend on life-cycle assessment (European Commission, DG XI, 1999). In the United States, the presidential order requiring green purchasing by Federal departments refers to environmental impacts throughout a product's life-cycle.

Why now?

Though the NIMBY[3] trend is still strong, times have changed. The horizon for environmental concern has widened. The little blue planet, seen from the moon for the first time 30 years ago, is a small world. The intensification of transportation networks, the development of the World Wide Web, but also the emergence of global environmental issues such as global warming, have paved the way to a new awareness of the environment. If the planet, and not only my 'back yard', is my environment, then I am concerned with the whole life-cycle of the products I consume, and my consumption choices can be guided by a label referring to this life-cycle. Diminishing space for garbage landfills has led to similar thinking within industry and government.

A new dividing-up of responsibility for health, risk and environment

Life-cycle thinking is a new way of taking responsibility for the effects of a product or process on society and the environment. In the old 'fordist' mode of production, responsibility for the product lay only within the factory. Bad effects further down the line were a sort of hot potato passed from one actor to the next. Thus, for example, car dismantlers were responsible for the toxic messes dumped on them, and the costs of cleaning were their problem.

3 Not In My Back Yard: this term signifies the refusal of citizens to live near polluted or polluting sites, but implies an apathy towards pollution issues elsewhere.

Today, with life-cycle thinking, those toxic junkyard problems become the responsibility of the manufacturers of the products that end up there. In Europe, Renault, BMW and Fiat are organising the collection and dismantling of automobiles that are no longer wanted. In the United States, USCAR, a consortium of US car manufacturers, grouped American auto-makers into a dismantling pilot project. 'Take-back' regulations are on the books in Europe, particularly for electronic products. New forms of co-operation among actors are taking shape.

Two major consequences are clear:

- Manufacturers, who are no longer isolated actors, are going to want to share the responsibility for their products with their supply chains.
- The consumer stage, which used to be of no interest to either the manufacturer or regulating bodies, becomes an essential part of life-cycle thinking and is likely to become even more so, as consumers become responsible for all the potential effects of the products they choose and use.

New modes of industrial co-operation

Exchange of information intensifies within sectors

In order to pinpoint responsibility for any given effect of a product, it is necessary to know the materials and substances used at each phase of manufacture. Industrial manufacturers now request specifications lists from their suppliers, who request specifications from *their* suppliers, and so on. In the case of DaimlerChrysler, in its life-cycle management programme it has created a new process called RSRC (Regulated Substance and Recyclability Certification). This process will inventory all the potentially dangerous materials and substances used in the manufacture of each vehicle so that the recyclability of each piece is possible. Such a process has obviously involved a huge effort and considerable motivation. All suppliers can then know the environmental impacts of their parts or processes.

Beyond the inventory described above, some product assemblers (original equipment manufacturers, or OEMs) are now requiring all suppliers to provide life-cycle assessments on all materials and parts provided. This is currently the practice in the German automobile industry.

In the case of Hewlett-Packard, its Procurement of Environmentally Responsible Material (PERM) allows for the evaluation of parts and processes at every phase of its operation. Suppliers are chosen on the basis of this PERM rating, which forces them to reduce to the minimum the environmental impacts of their parts if they wish to remain competitive (Choong 1996).

All operations of this sort imply complete exchanges of information between suppliers and manufacturers, and assemblers, all of which leads to more durable partnerships rather than opportunistic partnerships.

The evolution of product take-back

Companies now establish networks for dismantling, sorting, recycling and disposing of their discarded products (see Fig. 2).

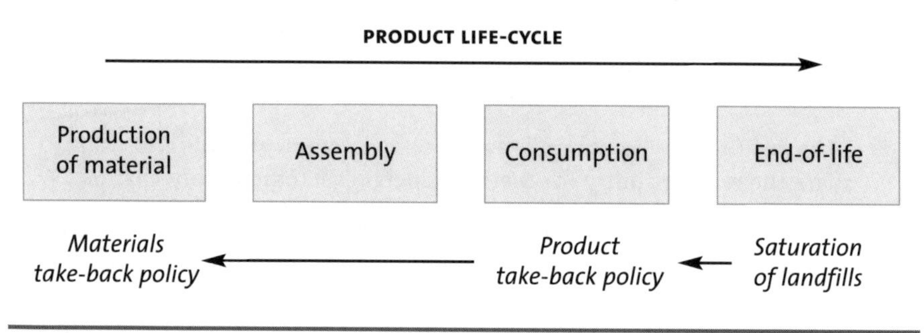

Figure 2 **Product life-cycle**

It is even possible to imagine, at least for simple products such as single-material packaging, that a supplier of such material remains the owner of the material forever, which will certainly encourage the creation of re-usable materials.

From products to services

Relations between consumers and manufacturers are redefined when we move from selling products to selling services. When manufacturers remain owners of the product and therefore responsible for its environmental impacts, they are motivated to truly implement life-cycle thinking. Reverse engineering, or reverse production systems, follow. Instead of the planned obsolescence of products, in order for the manufacturer to sell more and more, manufacturers will invest in creating sustainable products, which will automatically reduce the impact on the environment.

Toward responsible consumption

Whereas production is concentrated generally in identifiable industrial sites, consumption is spread out in time and space differently, depending on the product. It is therefore more difficult to deal with the potential pollution caused by individual consumption than with production pollution. Yet, especially with durable goods such as automobiles, more pollution occurs during use than during either manufacture or disposal. Much of this, of course, has to do with energy use.

In spite of the emergence of an ecologically concerned population, most Western consumers make no effort when it comes to the environmental effects of the products they buy. In some northern European countries, consumers have united to suppress the use of plastic bags in supermarkets, yet consumers in these countries tend to buy large cars with heavy fuel consumption and high emissions.

Regulations can play a role in regard to consumption. The Corporate Average Fuel Economy (CAFE) standards in the United States established a maximum average mileage per gallon for cars (across the whole fleet offered by each manufacturer) that manufacturers must meet. In France, regulations require an emissions test on all vehicles over three years old. Nevertheless, it is up to consumers to choose their cars, and both manufacturers and consumers sometimes prefer to pay fines rather than follow the rules. Most four-wheel-drive, off-road vehicles, all of them gas hogs, are currently a fad in the United States, and they are not regulated by the CAFE standards. Further, the manner in which one drives a car has an effect on the car's use of fuel. It may become useful for eco-conception (design for environment) to look into systems that control a car's use, and when this happens, marketing efforts will have to follow.

In order to further illustrate the importance of the consumer phase in the life of a product, let us look at food. Life-cycle assessments have been undertaken comparing paper and plastic packaging for bread. One such study shows a lower environmental impact for paper packaging, yet the consumer is going to throw out the bread that goes stale rapidly in the paper bag, and eat all of the still-fresh bread in the plastic bag. Obviously, consumer behaviour must be taken into consideration in the ecodesign of products. This is really a relatively unexplored field.

Conclusion: limits and further evolution of life-cycle thinking

Life-cycle thinking provides a rich framework of analysis and action for this era of increasing environmental constraints and competitive demands. Nevertheless, life-cycle assessment does not yet take into consideration certain externalities or interactions with neighbouring systems. For example, how can the noise levels in cities be taken into consideration? Or the gradual blackening of buildings as a result of passing traffic? Or illnesses resulting from pollution? No tool yet exists to include such things in the equation.

Further, in assessing genetically modified corn, how can we measure effects of transmitting to a neighbouring organic cornfield a gene resistant to the only antibiotic allowed in organic farming? On a case-by-case basis, life-cycle thinking must extend the assessment to include neighbouring products or infrastructures—or else we must acknowledge the temporary and partial nature of its conclusions. It is this sort of consideration for neighbouring industrial systems that allows for zero-emission or industrial symbiosis zones.

Generally speaking, life-cycle assessment must look beyond the product under consideration to an assessment of the service provided, the need fulfilled by the product in order to find truly radical long-term solutions. The best solution might have nothing to do with the original product at all. It wasn't through trying to lessen the environmental impact of writing paper that e-mail came about, for example, yet the environment has been the beneficiary.

Editor's note. As an example of 'life-cycle thinking' moving into the mainstream, in February 2000 the European Parliament approved rules that force car manufacturers to bear the cost of recycling cars in the 15 countries of the European Union. National capitals will translate the rules into laws in each country. The 'End-of-Life Vehicles Directive' established that, starting in 2006, manufacturers will have to take back and recycle all cars made even before the legislation was created. The directive set a precedent of making companies responsible for their products 'from the cradle to the grave'.

ECO-EFFICIENCY

What is eco-efficiency? Eco-efficiency combines environmental and economic performance to create more value with less impact. At the micro level, eco-efficiency is mainly a business concept. At the macro policy level, it can describe important elements of the move toward sustainable development. The European Partners for the Environment (a group of both businesses and non-governmental organisations) say that, in order to become eco-efficient, companies need to:

- Reduce the material intensity of their goods and services
- Reduce the energy intensity
- Reduce the dispersion of toxic substances
- Enhance material recyclability
- Maximise the sustainable use of renewable resources
- Extend material durability
- Increase the service intensity of their goods and services

In addition, businesses must also use creativity to deploy new technologies, initiate improvements along the supply chain, and bring new products and services to market.

Following are a series of texts further elaborating eco-efficiency, the first of which comes from David Gee of the European Environment Agency (EEA). The EEA is part of the European Commission, the government body for 15 European countries. The EEA acts on behalf of the European Commission in regard to environmental matters, and the Agency's work is assessed in light of the Commission's goals. The European Environment Agency mission, as described in European Commission Regulation 1210/90, adopted in May 1990, is to help achieve significant and measurable improvement in Europe's environment through the provision of timely, targeted, relevant and reliable information to policy-making agents and the public.

'Meeting needs, consuming resources' is taken from the EEA's report, *Environment in the European Union at the Turn of the Century* (EEA 1999a). Before joining the EEA, David Gee worked with trade unions, with Friends of the Earth, and as a partner of WBMG Environmental Communications in Great Britain. He is an example of what some people call a 'policy wonk', and the following excerpts speak with a distinctly 'agency' voice.

MEETING NEEDS, CONSUMING RESOURCES

David Gee

THE PLANET IS AN INTEGRATED SYSTEM OF ENERGY AND MATERIAL FLOWS which involves the circulation of carbon, chlorine, nitrogen, sulphur, water and other key elements between the environmental compartments of air, water, soil and vegetation. The sun is the initial driving force behind such activity. This environmental system not only sustains individual life via air, food and drink but also enables us to collectively organise food, clothing and shelter in an economic subsystem through the provision of:

- Sources of energy and materials
- Sinks for waste and pollution
- Services such as water flow regulation
- Space for people, nature and aesthetics

These four basic 'life-support' functions of the environment are essential to any economy, but, while the products of nature such as food and drinking water are vital, the more hidden, but essential, ecological services are often ignored, or undervalued. For example, rivers and wetlands not only provide fish, water and facilities for recreation but scientific advances show that their servicing functions include holding and circulating water, producing oxygen, storing carbon dioxide, helping to regulate climate, and filtering pollution.

Box 1 **Economies depend on the environment**

Ecological services, unlike human-made technologies, are largely free, but their value can depreciate, and may disappear with over-use, as in the case of energy and materials taken from the environment, converted into useful products, then returned to the environment as waste and emissions. Such 'economic metabolism', if it exceeds the resilience of the environment, could cause shortages of both resources and ecological services.

However, managing the exploitation of the *sources* of energy and materials from nature, such as metals, minerals and forests, is much easier than managing the ecological *services* of nature, such as climate regulation, nutrient recycling, waste assimilation, and radiation protection from the ozone layer.

Shortages of materials can be overcome by improvements in efficiency, or via alternative products, such as plastics from biomass waste. Furthermore, the deposits of metals and fossil fuels are usually owned by someone, so that control over their use, via price and other means, is possible. Scarcity, and its associated price rises, stimulates invention, and human-made capital can sometimes replace natural materials from nature.

Ecological services are more difficult to deal with. It is not possible to replace the ozone layer or the climate regulatory systems, with human-made capital, and their efficient functioning can fail once thresholds of 'load' are passed. Such ecological services are not owned by anyone, nor do they usually have prices, so preserving them via market mechanisms is not so easy.

It is therefore concern about the current systems of economic activity overwhelming the *sinks* and destroying the *services* from the environment, rather than possible shortages of energy or materials, that have moved scientists, politicians and others to suggest that radical change in the way that we meet our needs is required.

Eco-efficiency: getting more from less

Meeting needs with less use of natural and human-made resources but with more use of people has become an environmental and economic imperative since 1993 (see Box 2). 'Eco-efficiency' aims at decoupling resource use and pollutant release from economic activity and is becoming an object of environmental policy (OECD 1998a; EEA 1998b).

“*The serious economic and social problems the Community currently faces are the result of some fundamental inefficiencies: an 'under-use' of the quality and quantity of the labour force, combined with an 'over-use' of natural and environmental resources . . . The basic challenge of a new economic development model is to reverse the present negative relationship between environmental conditions and the quality of life in general, on the one hand, and economic prosperity, on the other hand*” (European Commission 1993).

Box 2 **Less nature, more people?**

The Agenda 21 update (1997), in its paragraph on integration, notes the need to improve the efficiency of resource use; to consider a tenfold improvement in resource productivity in industrialised countries; and to promote measures favouring eco-efficiency. This will require breaking the links between use of nature, as measured by environmental indicators, and economic development, as measured by output indicators, such as gross domestic product (GDP), or passenger-kilometres in transport, for example. Both 'use of nature' and 'welfare' indicators need improving in order to better reflect reality and human needs, but some current trends in eco-efficiency can be gauged from using existing information.

Improved eco-efficiency is not a sufficient condition for sustainable development, as absolute reductions in the use of nature, and associated environmental pressures, may be necessary to get within the Earth's (and human) carrying capacities, so that both relative and absolute de-linking between the use of nature and economic growth will be necessary.

Figure 3 summarises progress with the de-linking of some environmental indicators from economic growth in the EU in the first half of the 1990s, with outlooks to 2010.

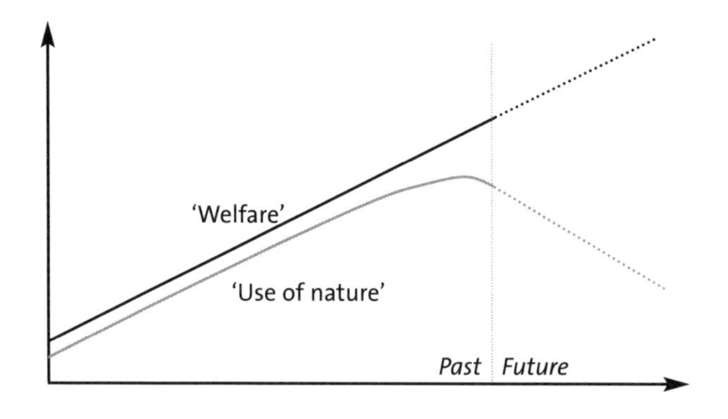

Figure 3 **Relative and absolute de-linking**

Source: European Environment Agency

The case of Austria, which was the first country to adopt the Factor 10 target in its national environmental plan, illustrates the difference between relative eco-efficiency gains and the continued rise in the absolute use of resources from economic growth (Fig. 4).

There are two broad ways to enhance eco-efficiency:

- Via the more elegant and equitable use of resources, through innovation in the use of resources and labour
- Via a focus on meeting human needs more from labour-intensive services than from capital-intensive products

There is considerable potential for initiatives by firms and communities to improve eco-efficiency using current technologies. For example, manufacturers have found profitable ways to reduce their use of materials, energy and water per unit of production by 10%–40% (OECD 1998a) and initiatives in the services sector, local governments and households achieve similar savings. Firms have also demonstrated technologies that cut the use or emission of toxic substances

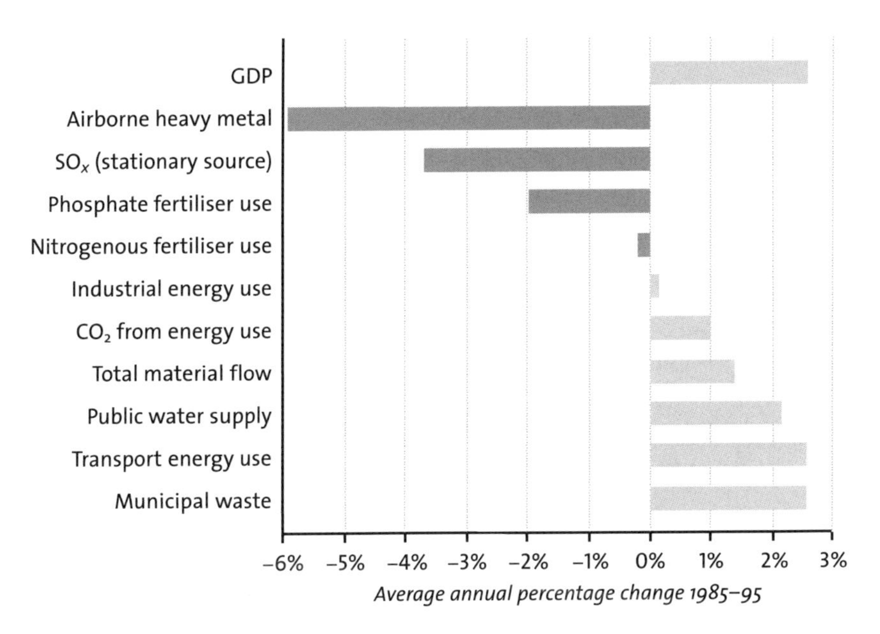

GDP = gross domestic product; SO_x = oxides of sulphur; CO_2 = carbon dioxide

Figure 4 **Eco-efficiency and material flows in Austria**

Source: Schuster 1997

by 90% or more, although these technologies are not always put into place (OECD 1998a; von Weizsäcker *et al.* 1997). A few firms have taken initiatives to reduce environmental impacts during and after the use of products: for example, by recovering used equipment and re-using durable components. Initiatives that address impacts over the full life-cycle offer the greatest potential for reducing pollution and resource use economy-wide, but few firms have developed comprehensive strategies for achieving this. Business organisations such as the World Business Council for Sustainable Development (WBCSD) are encouraging reductions in the intensity of energy and materials use via the promotion of eco-efficiency (Box 3). 'Demand-side management' in the energy, water, transport and parts of the chemicals sector is beginning to shift the focus from consuming products to using services, with associated eco-efficiency and employment gains.

Industrial ecology has been slowly emerging as an approach to eco-efficiency and sustainability since the early 1970s (Erkman 1997). It includes the promotion of regional recycling networks (or industrial ecosystems) such as the industrial symbiosis networks in Kalundborg, Denmark, parts of the Ruhr, Germany, and Styria, Austria, which already involve using the outputs of substantial quantities of waste from some companies as inputs for other companies. For example, of the estimated 3.8 million tonnes of non-construction waste generated each year

1. Minimise the material intensity of goods and services
2. Minimise the energy intensity of goods and services
3. Minimise toxic dispersion
4. Enhance material recyclability
5. Maximise the use of renewable resources
6. Extend product durability
7. Increase the service intensity of goods and services

Source: WBCSD/EPE 1999

The materials intensity of two different types of kitchen illustrate the application of some of these criteria (Fig. 5).

Box 3 **Eco-efficiency criteria of the World Business Council for Sustainable Development, and their application to a kitchen**

Figure 5 **Materials intensities: kitchens example**

Source: Liedtke *et al.* 1994

in Styria, about 1.5 million is now used as production inputs to iron manufacturing, construction materials, paper and cement plants within the recycling network (Schwarz and Steininger 1997).

Eco-industrial parks (Lowe 1997) are being developed, mainly in the USA and Japan, where the principles of industrial symbiosis and 'zero emissions' (Pauli 1997) are being designed into the development plans of the parks. Although there are thermodynamic, energy and economic limits to recycling, the current high ratio of wastes to useful products indicates that there is considerable scope for the more efficient use of resources.

The search for innovative chemical processes that facilitate less toxic and resource-intensive chemical production (Box 4) is being stimulated by 'green chemistry' networks in Germany, Italy, the UK, Japan and the USA (Anastas and Breen 1997; Tundo and Breen 1999). As the US Academy of Engineering has pointed out, 'design should not merely meet environmental regulations: environmental elegance should be part of the culture of engineering education' (Jackson 1996). Those companies and countries that first succeed in emulating nature's elegance in resource use will provide a great service to the environment and human society (EEA/UNEP 1998).

In general, the focus on eco-efficiency will lead to the development of circular, rather than linear economies, where wastes become inputs rather than outputs.

- Clean synthesis (e.g. new routes to important chemical intermediates including heterocycles)
- Enhanced atom utilisation (e.g. more efficient methods of bromination)
- Replacement of stoichiometric reagents (e.g. catalytic oxidations using air as the only consumable source of oxygen)
- New solvents and reaction media (e.g. use of supercritical fluids and reactions in ionic liquids)
- Water-based processes and products (e.g. organic reactions in high-temperature water)
- Replacements for hazardous reagents (e.g. the use of solid acids as replacements for traditional corrosive acids)
- Intensive processing (e.g. the use of spinning disc reactors)
- Novel separation technologies (e.g. the use of novel biphasic systems such as those involving a fluorous phase)
- Alternative feedstocks (e.g. the use of plant-derived products as raw materials for the chemical industry)
- New safer chemicals and materials (e.g. new natural product-derived pesticides)
- Waste minimisation and reduction (e.g. applying the principles of atom utilisation and the use of selective catalysts)

Box 4 **Green chemistry: key objectives**

Source: Green Chemistry 1.1 (February 1999)

THE WBCSD'S WORKING GROUP 'ECO-EFFICIENCY METRICS AND REPORTING' recommends using the following ratio as a general equation to measure and report eco-efficiency:

eco-efficiency = unit of value provided per unit of environmental burden

The following cross-comparable indicators have been considered by the WBCSD working group:

Environmental indicators

- Total amount of energy use
- Total amount of materials use
- Greenhouse gas emissions
- Ozone-depleting substances emissions
- SO_2 and NO_x emissions

Value indicators

- Mass or number of product
- Number of employees
- Sales/turnover
- Gross margin
- Value added

Box 5 **Corporate reporting on eco-efficiency**

Source: World Business Council for Sustainable Development, *Executive Brief*, January 1999

The Organisation for Economic Co-operation and Development (OECD) has identified several ways in which governments could encourage eco-efficiency initiatives by firms and communities, such as: tax and subsidy reform; regulations; promoting 'extended producer responsibility'; and supporting the development of standard monitoring and reporting procedures.

ECO-EFFICIENCY
The chemicals industry

To illustrate the idea of eco-efficiency in one particular corporation, we offer the following excerpts from a talk given in Paris at ECO 1997, an international conference on eco-strategies, by David Buzzelli. Buzzelli was at the time co-chair of the US Presidential Advisory Commission on Sustainable Development and the Dow Chemical Company's Vice President for Environment, Health and Safety. For many people, the name Dow remains forever attached to the napalm the company manufactured during the Vietnam war. Yet, today, Dow is engaged in continuous improvement in environmental performance, a commitment sustained by long-term management support.

The chemical industry's boom and bust cycles have created tremendous pressure for companies to consolidate around their core competences. In the mid-1990s, Dow briefly invested in an environmental division, hoping to make money in environmental consulting, but this was dropped along with the pharmaceutical division when it became clear that such diversification was not a sound strategy. For example, research is crucial in pharmacy, where it can cost $100 million to develop a new molecule and thus a new drug—too costly for Dow to be competitive. Dow merged with Union Carbide to become number two worldwide (after DuPont) in basic chemical (petroleum-based) products. Nevertheless, Dow's commitment to continuous environmental improvement remains in place. Their *Public Report 1999* is an interesting model in environmental/social reporting. Included in it are goals for the year 2005 to further reduce energy use and air and water emissions for global operations as follows:

- Priority compounds by 75% (priority compounds include persistent, toxic and bioaccumulative compounds, known human carcinogens, selected ozone-depleting substances, and high-volume toxic compounds)
- Chemical emissions by 50%
- Energy use per pound of production by 20%

In the following, David Buzzelli describes Dow's move from the principles developed by the Business Council for Sustainable Development to their concrete implementation in a major chemical company. (See also Part 4, 'Civic actions for change', for a 1999 view involving the Dow Chemical Company and the Natural Resources Defense Council: 'A daring partnership pays off', page 278.)

THE CHALLENGE OF ECO-EFFICIENCY

David Buzzelli

The chemical industry and the environment

A fundamental change has taken place in the chemical industry's attitude towards the environment over the last 35 years. From managing single products and plants, the industry has moved to managing complete product life-cycles.

In 1989, Dow's Annual Report included for the first time a special shareholder's report on the environment. The following statement which appeared on the cover illustrates how environmental policy was evolving towards greater integration with business performance:

> One issue, more than any other, will affect Dow's prospects in the 1990s and beyond. That issue is the environment.

Fundamental change in society and human needs, coupled with a significant increase in global population, will alter existing consumption patterns.

The sustainable corporation must transform materials and provide services that people value for the contribution to the quality of life and the protection of the environment. This means that the goods and services provided can be consumed by an increasingly large segment of the population to respond to population growth and the need for more equity. This consumption growth should not threaten the ecological security.

Production, supply and disposal systems must be designed and operated in a way that reconstitutes and maintains the environmental quality. This requires cleaner technologies and developments such as alternative non-fossil energy and raw material sources, and integrated processes or closed material loops.

Eco-efficiency is broadly defined as the production, delivery and use of competitively priced goods and services, coupled with the achievement of environmental and social goals. The Business Council for Sustainable Development

(BCSD), in its 1992 publication, *Changing Course* (Schmidheiny 1992), proposed the following definition for eco-efficiency:

> Eco-efficiency is reached by the delivery of competitively priced goods and services that satisfy human needs and bring quality of life while progressively reducing ecological impacts and resource intensity, through the life-cycle, to a level at least equal with the Earth's estimated carrying capacity.

1. An eco-efficient product is durable, repairable and re-usable, and therefore more attractive to consumers.
2. An eco-efficient business takes account of its environmental responsibilities when designing technologies, processes and products. In doing so, it finds opportunities for efficiency gains and other overhead savings.
3. Eco-efficiency is not an absolute. The notion will evolve as a function of innovation, customer values and economical policy instruments. It represents the direction of an effort.
4. At Dow we have developed a six-dimension model that we call the Eco-Innovation Compass. This Compass helps us to evaluate existing products and guide us in the development of new or improved ones. It can play a role in helping us incorporate the concept of eco-efficiency into our business strategies by providing insight into how we can make environmentally improved products that are also commercially viable.

Before evaluating any product, a company must first change its perspective and look beyond the end-use product. One must consider the total design–make–supply–use system and the total life-cycle of the product from initial raw materials to final waste products after use. Eco-efficiency must be evaluated and improved in this context of the life-cycle of a total system. This implies that one must look at the function fulfilled by the product rather than the product itself.

The Eco-Innovation Compass compels us to scrutinise six dimensions of the total system:

1. **Mass:** the total of raw materials, fuels and utilities consumed in the system during the life-cycle to deliver the desired function. The opportunity is to significantly reduce the mass burdens and dematerialise the way the system provides quality of life and benefits to the market chain.
2. **Energy.** The opportunity is to spot the parts of the system and the life-cycle that have the highest energy intensity and redesign the product or its use to provide significant energy savings.
3. **Environmental quality and human health.** Reduce and control the dispersion of elements that have negative environmental or health impacts when they reach, or accumulate to, a level beyond a critical dose for the environment or humans.

4. **Material utilisation.** Designing for recyclability is important; recycling effectively and efficiently is even more important. Another opportunity is to design the system as part of a larger natural cycle. Materials are borrowed and returned to nature without negatively affecting the balance of the cycle.

5. **Renewable materials.** In some cases, these materials have advantages over reactive chemistry from a total cradle-to-grave perspective.

6. **Durability and functionality.** Extending the durability and service part of a system, especially in the usage phase, can improve eco-efficiency. Improving the functionality of products also increases their eco-efficiency (e.g. Swiss army knife).

Getting eco-efficient is a matter of redesigning a system in every possible respect. One must consider reductions in mass and energy utilisation, fewer toxic chemicals, improvements in recycling or in the use of renewable resources, and innovation in service life and functionality. The six dimensions are not independent. They overlap and interrelate significantly.

And it works—here are some examples:

1. Dow developed 'closed-loop' systems for collecting, recycling and ultimately reselling chlorinated solvents used in the dry-cleaning and metal-cleaning industries throughout most European countries. Closing the loop on products reduces the net environmental impact of our products and processes, while satisfying the needs of a growing market. The programme was pioneered by Dow affiliate Safechem Umwelt GmbH in Germany, Austria and Switzerland. (This is a 'product-to-services' shift; see the Manzini paper on page 77.)

2. DowElanco developed a new technology to control termites (Sentricon Colony Elimination System) which is more effective, less toxic, less odorous and requires substantially fewer material inputs than most other methods.

3. Due to increasing public concern about toxic materials in televisions, Sony Europe developed the 'Green TV' which is lighter, less energy-intensive and creates less health and environmental risk than previous models. It also has better picture quality and is cheaper to produce than its predecessor.

4. Dow developed a nanofiltration membrane, FILMTEC NF200B, for surface water potabiliation for SEDIF (Syndical d'Eau de l'Ile de France) which removes organic components below the legal limits and limits all possible food sources for bacteria while improving the overall quality of the drinking water.

All these examples show us that the name of the game is significant change, not incremental change.

ECO-EFFICIENCY
The electronics sector

Kyehwan Oh is a senior executive vice president and head of the semiconductor sector of Hyundai Electronics Industries Co. Ltd (HEI). He is chairman of the environmental management committee in HEI. Dr Oh received his BA in applied physics from Seoul National University, and his PhD in physics from Iowa State University in the United States. He has over 20 years of semiconductor industry experience as an engineer and manager. Prior to joining HEI, he worked at AT&T Bell Labs in New Jersey for seven years and at Hewlett-Packard in California for a year. He holds several patents such as the techniques for Integrated Circuit Contact and for Doping from Polysilicon Transfer Layer.

Dr Oh delivered the following paper in Paris at ECO 1999, an international conference on eco-competitiveness. He underlined the newness of the semiconductor industry, reminding everyone that the transistor was invented only 50 years ago. The semiconductor was a 0.2 μ technology in 1999,[4] the average human hair being 100 μ wide. So the technology deals with objects one-five-hundredth of the width of a hair. Now, of course, we are on the verge of a new technology in semiconductors where components may be only one molecule wide. This will lead to a discontinuity, a huge technological leap that will completely alter the electronics industry.

In the current technology, the cleaning of semiconductor wafers is critical and uses highly toxic chemicals and gases such as sulphuric acid and hydrochloric acid. Dr Oh explained wryly that it was when Hyundai Electronics built a factory in Eugene, Oregon, a city known in the United States for its large number of environmentalists, that HEI was forced to develop many of the extreme safety measures the group now practises in all its plants. When asked if he thought HEI had made a mistake in choosing Eugene, he said that, while he couldn't quite believe some of the extraordinary demands made at the time, he felt they'd been kicked into the future and was grateful for it.

Dr Oh also said that travelling to Paris to present his paper was the first time that he, and Hyundai Electronics, had spoken outside Korea about HEI's environmental practices, and that he was quite uncertain how they might measure up. In any case, contrary to the idea that the new information technology exists 'in thin air' and therefore has no environmental consequences, its manufacture involves enormous amounts of energy and highly toxic chemicals.

4 A micron (μ) is a metre × 10^{-6}.

MANAGEMENT OF CHEMICALS IN THE MICROELECTRONICS ENVIRONMENT

Kyehwan Oh

Hyundai Electronics was founded in 1983 to produce semiconductor, multimedia, communication and industrial electronics products. Hyundai Electronics has grown into a global company with seven production subsidiaries and seven sales subsidiaries in nine countries, resulting in sales of $3.2 billion, and with over 15,000 employees in 1998.

From the environmental perspective, our base company is strictly controlled by governmental regulations, because it is located near a water resources preservation zone, which provides drinking water for more than 20 million people. We operate our business in the 'critical condition' that an environmental accident would have a serious impact on all our business activities. On the other hand, this critical condition has led to environmental issues being considered as a major part of business management and, in 1994, we implemented environmental management activities in which all employees participate.

Through 1998, the pollutant discharge rate per product unit has been reduced by over 20% every year since 1994, and, externally, we were certified to BS 7750 and ISO 14001, the environmental management system standard. We were declared an 'environmentally friendly company' which is similar to the EMAS (Eco-management and Audit Scheme) regulation. Finally, in 1998 we won a prize as 'the model company in environmental management' from the President of Korea in recognition of our efforts. Thus, my company has been recognised in Korea as one of the leading companies in environmental management.

This performance resulted from the particular location condition of Hyundai Electronics and the rapid change of management circumstances in the 1990s.

Since 1990, there has been a great deal of public debate on pollution prevention technology. In Korea, concern about environmental pollution increased, environmental regulations were strengthened, and residents, stockholders and customers became interested in companies' environmental management. These internal and external pressures accelerated the environmental management activities of Hyundai Electronics.

In introducing environmental management, we began by documenting every environmental impact resulting from business activities, products and service. We created 11 categories, such as R&D, purchasing, manufacturing and so on, and evaluated the environmental impacts of each. Following this evaluation, we obtained environmental information for the complete life-cycle of our products, from manufacturing to disposal. This showed us the huge environmental impacts of our internal manufacturing processes. Most notably, the manufacturing process seriously affects global warming and ecological toxicity, because various chemicals, gases and a great deal of electricity are used in the manufacturing process. Therefore, two categories—chemicals and electricity use—were selected as objectives for improvement, and they have improved continuously.

In order to control the environmental impacts caused by electricity usage, we constructed two co-generation power plants and began to use electricity more efficiently. Also, we have established and implemented an annual energy-saving programme.

To control the environmental impacts of chemical usage, we have implemented the following three activities in which management and staff participate consistently:

- Reducing chemical usage
- Minimising the possibility of chemical accidents
- Treating waste chemicals in safe and stable ways

Detailed targets and programmes for achieving the three major objectives were established in all functions, such as R&D, the wafer fabrication process, the package process, utility operation, and waste treatment. In addition, the CEO of Hyundai Electronics showed strong leadership by insisting that all employees take part in achieving the targets, and by providing necessary training.

Reducing chemical usage

Because reducing chemical usage is the best way to solve most environmental impacts, we treated this objective as our highest priority. We implemented this objective in three ways.

The first way was to develop new processes, unify process steps or shorten processes. The R&D and engineering departments developed technology that minimises process steps and changes the existing process to a non-chemical process, while reviewing existing processes and experimenting and testing. We developed processes to avoid cleaning by nitric acid (HNO_3) and acetic acid (CH_3COOH), and shortened processes using hydrochloric acid (HCl) from two-step processes to one-step. These are two good examples.

The second way was to operate processes with the minimum amount of chemicals by process condition optimisation. We achieved this by conducting many experiments to optimise key parameters such as changing the time of the chemical bath, dispensing time and frequency.

For example, the process recipe of the developer-input rate was changed from 107.55 cc per product unit to 72.7 cc per product unit. This improvement led to a reduction in the annual usage of developer by 224,400 l. And the input rate of photo resistor was changed from 5.6 cc per product unit to 2.7 cc per product unit, leading to reduced annual usage for photo resistor of about 4,500 l.

The third way was to recycle discharged waste chemicals. We redesigned the discharging line to separate waste chemicals by type and introduced sulphuric acid (H_2SO_4) and isopropyl alcohol (IPA) reprocessing facilities while reviewing technology and economics. The sulphuric acid reprocessing system makes pure sulphuric acid from impure sulphuric acid by heating it to boiling point. We have thus recycled 70% of our sulphuric acid usage and reduced purchasing costs by $2.5 million per year. We also recycle 30% of IPA usage by using an IPA reprocessing system based on osmotic action. This has reduced purchasing costs by $0.8 million per year

As a result of all these activities, toxic chemical usage per product unit has declined 23% every year since 1994.

Chemical accident prevention

We have formulated an accident prevention system for chemical leakage and spillage with chemical usage reduction.

When we procure chemicals, we purchase the chemical with the least environmental impact and reduce the chances of an accident from transportation and handling of chemicals by means of a double-packed case.

Rainwater lines surrounding chemical storage and handling places are enclosed, and all trenches are connected to waste-water treatment facilities. A dike for prevention of chemical spillage is installed at each chemical tank. Also, we minimised the manual operations in chemical supply processes by automation. Lines for toxic and chemical supply are installed using a double protection system in case of accidental dispersion of toxic gases.

Software aspects in accident prevention were considered together with hardware ones. Compulsory training for employees who handle chemicals is conducted for two hours every month. We evaluated the control level and prevention facilities for areas where chemicals are transported, stored and used so that we can grade them for degree of hazard and prepare appropriate countermeasures. Operational procedures are prepared for accident prevention in vulnerable areas with specific emergency plans. Through regular emergency disciplines, accidents can be prevented and minimised with rapid responses.

As a result of these activities, we have achieved zero accidents from chemical handling, storage and supply activities since 1995.

Treating waste chemicals in safe and stable ways

The last area we concentrated on was the safe and stable treatment of waste chemicals by improving each facility for maximum efficiency.

In waste-water treatment, separate lines for hydrogen fluoride, acid and solvent waste-water were installed to allow for characterised treatment of the waste-water. This improved treatment efficiency with the least chemical usage. Also we operated nitrogen-removing facilities in waste-water before any other company in Korea, and the nitrogen concentration in the effluent which causes eutrophication in water has therefore reduced by 55% since 1994.

In air treatment, the discharging gases with acid, toxic, solvent and general characteristics are being treated with special scrubbers. The pollutant concentrations are controlled to 10% below the legal requirement. Moreover, for volatile organic compound (VOC) treatment on-site, we have installed and operated facilities in advance of legislation.

We have set up internal standards that are 50% below the legal requirements of pollutants discharged outside, and the compliance of emission and effluent has been monitored on a real-time basis by a telemonitoring system.

For waste chemicals collected and treated outside, the qualifications of the outside contractors are reinforced with a periodical monitoring system for the appropriateness of treatment.

Business benefits of environmental management

We were able to achieve both visible and invisible performance improvements through the chemical management programme initiated in 1994. From process improvements and recycling, we have saved $22.5 million annually since 1995, and the development of appropriate treatment technologies for pollutants has made it possible for the concentration of effluent to be maintained sufficiently low so that we do not need to pay unnecessary pollutant discharging fees of several tens of millions of dollars when securing business.

From the environmental perspective, we have accomplished unprecedented performances so that pollution with respect to semiconductor production has reduced rapidly, and pollutants from air, water, and toxic waste chemicals have dropped by 14%, 31% and 23%, respectively every year since 1995.

Future focus and conclusion

In the electronics industry, especially the semiconductor industry, I believe that one of the best ways to attain improvements in profitability while conserving the environment is to concentrate on source reduction of chemicals. We, Hyundai

Electronics, will consider the environmental impacts at every stage of product development and marketing, based on environment management know-how acquired. And, by the year 2000, the amount of waste toxic chemicals and water pollutants will have decreased by 26% and 30%, respectively, compared to 1998, while air pollutants discharged will have been kept to the same level as 1997 by continuous process improvements. For the scientific and systematic evaluation of all the processes of semiconductor manufacturing, we are introducing life-cycle assessment which will advance our environment management activities one step further. Also, we are trying to develop environmental technologies on recycling and pollutant treatments for distribution to other industries.

Many scientists predict the 21st century will be the century of the environment. Domestic and overseas environmental requirements are becoming more and more diverse and strict due to serious global pollution. We cannot compete any longer without considering the environment.

In light of this, we think the timing for our environmental activities focused on chemicals was appropriate. Moreover, we believe we can get valuable performance improvements such as a reduction in cost with environmental improvements at the same time.

ECO-EFFICIENCY
Waste-free: remanufacturing

Is 'waste-free' a metaphor? Or is it a slogan? In the early 1990s, Xerox Corporation made a commitment to become a waste-free company, leading to programmes that reduce waste during manufacture, use and at the end-of-life of products. Xerox has now become a leader in remanufacturing, an industrial process in which a discarded product is disassembled, its re-usable parts cleaned, refurbished and put into new products. Such labour-intensive remanufacturing uses one-fifth of the energy and one-tenth of the raw materials needed to manufacture a totally new product.

Xerox now builds remanufacturing capability into the new products it designs and delivers, and 90% of Xerox-designed equipment is remanufacturable. In 1998, Xerox re-used and recycled more than 145 million lb of material. With dematerialisation a goal on the horizon (no more heavy machines in the office, but telematic access to centralised machines), Xerox promises to leverage the benefits of information technology to further minimise the use of valuable resources. Perhaps **'dematerialisation'** will turn out to be the truly dynamic metaphor from Xerox.

Jack Azar is Vice President, Environment, Health and Safety of Xerox Corporation, and is therefore responsible for policy and strategy development and strategic implementation of all EH&S programmes at Xerox. After earning his PhD in Organic Chemistry from Columbia University, Azar joined Xerox as a materials technologist and then moved to management. He became Vice President in 1997. He is the holder of seven US patents and has co-authored many publications related to product design for environment. A native New Yorker with an ample sense of humour, Azar presented the following in Paris in 1999, using Powerpoint to show images, while he improvised from the knowledge he carries around in his head.

XEROX
Environmental leadership programme

Jack Azar

Xerox is a world leader in providing value to its customers, employees and society in all areas of environment, health and safety. Our desire is to obtain 100% customer environmental satisfaction. Through the Environmental Leadership Programme we have developed partnerships with government, with environmental groups, and with suppliers, the latter based on the new extended-enterprise concept in which Xerox works co-operatively and actively with our entire supply chain, often an exciting challenge in far corners of the world.

Figure 6 is not really an organisation chart, but a group of projects spread throughout Xerox operations that are brought together under the Environmental Leadership Programme. The steering committee with representation from all sectors provides guidance to the overall programme.

The cartridge-recycling programme and equipment remanufacturing were the first programmes to be implemented in 1991. More recently, the environmental products concept broadened our commitment considerably. Environmental marketing was established recently to proactively support the acceptance in the marketplace of remanufactured products, as customers initially felt that there was something inferior about remanufactured products, which is not the case at all. Currently, Xerox's environmental policy emphasises worldwide adoption of all programmes, an enormous undertaking.

Xerox's **waste-free** goal is a simple concept, one that is good for Xerox, for our customers, and for the environment. Waste-free does not mean zero waste, but rather a 90% reduction of emissions and waste from a baseline year (1990). The goal is to have waste-free products manufactured in waste-free factories to enable our customers to achieve waste-free offices. In the designing and building phase, this involves the efficient use of energy and materials, low emissions and noise, the minimal use of hazardous substances, and maximised parts re-use and

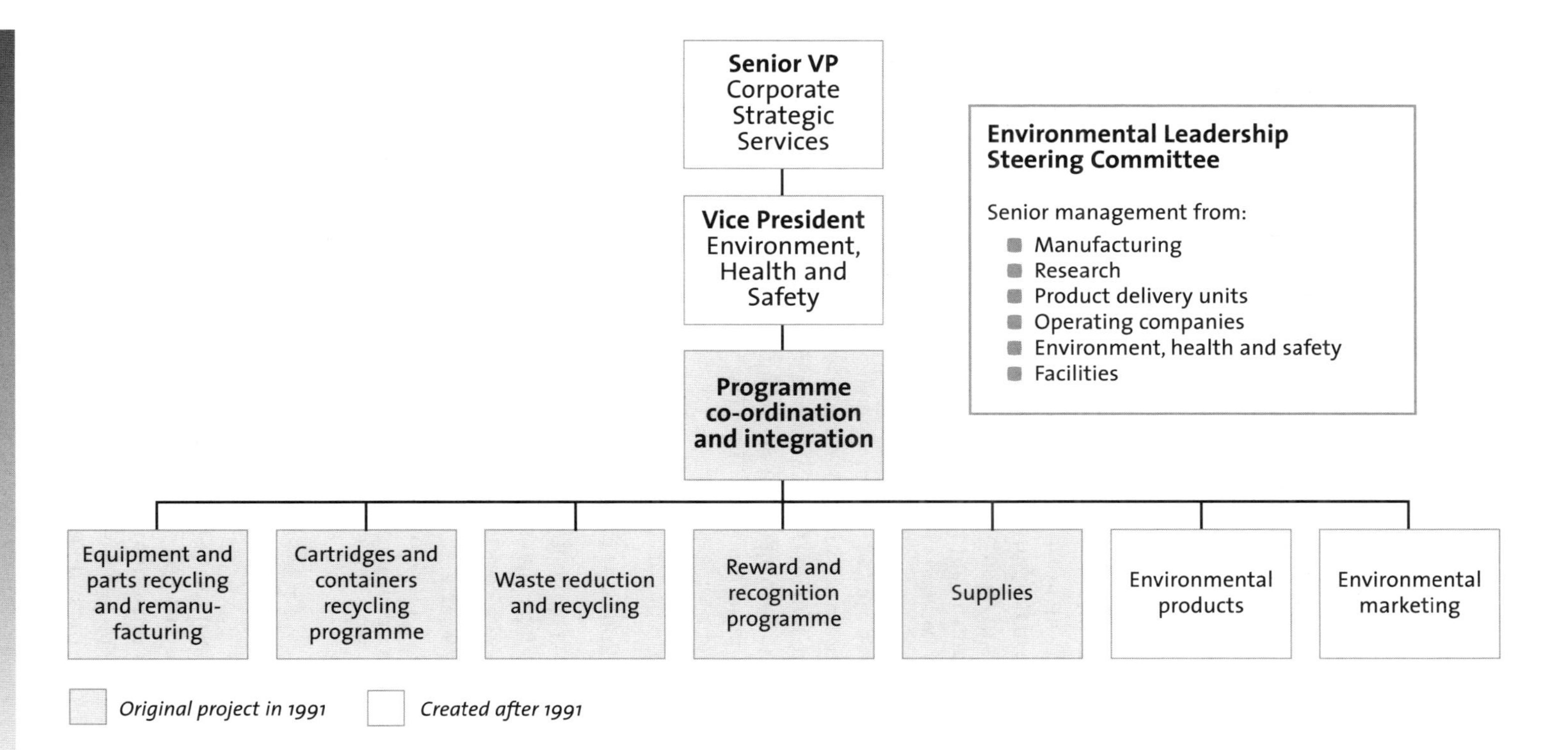

Figure 6 **Environmental Leadership Programme**

materials recycling. In the customer-use phase, this involves paper-saving features, energy efficiency, and returns programmes. In end-of-life management, we have equipment take-back followed by parts re-use and recycling, in which the smallest possible loops are both economically and environmentally preferable. And equipment is remanufactured into **original,** not used products.

Waste-free products

The drivers of waste-free products include:

- Product take-back regulations. Various European countries have take-back regulations in place; there is a European Directive on Waste Electrical and Electronic Equipment; there is a recycling law in place in Japan, etc.
- Environmental label certification criteria now address energy efficiency, product recycling capability, emissions limits and ISO 14001 certification.
- There is a growing acceptance of product recycling by the public.
- Customers increasingly require noise, heat and ozone reductions, plus waste return.

Our first waste-free products were our copy and print cartridges. The earlier designs were not recyclable—they were for single use only. At the beginning, we had a limited recovery of parts and very little recycling of plastic material.

Now, our new copy cartridge designs are our first example of 'design for environment'. They can easily be remanufactured. They are designed for disassembly, with many parts designed for multiple customer lives, with recyclable plastic materials, with price discounts as incentives for customer returns, and with lower life-cycle costs, which benefit both Xerox and the customer. Depending on the region, copy and print cartridges are returned to us by UPS, by post or by a special Xerox service. Our worldwide return rate is the highest in the industry. Beginning in 1995, 60% of cartridges were returned for re-use in the US and Canada, and 47% in Europe. In 1998, 64% were returned for re-use in US and Canada, and 74% in Europe.

Figure 7 gives us the picture on toner-container return and Figure 8 gives that for toner return, after which some is remanufactured and some recycled into non-Xerox uses. Our research programmes are actively seeking to develop other uses for returned toner.

The results of all supplies-return programmes in 1998 are that re-use and recycling of returned supplies diverted 9.5 million lb of material from landfill, including 595,000 lb of waste toner, 1,300,000 lb of retail cartridges, 1,940,000 lb of toner containers, and 5,610,000 lb of office cartridges.

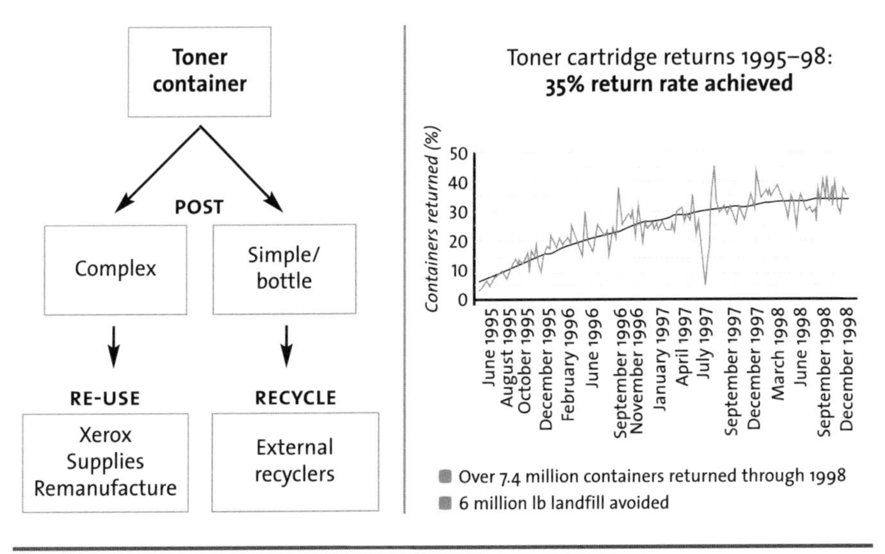

Figure 7 **Waste-free products: toner-container return**

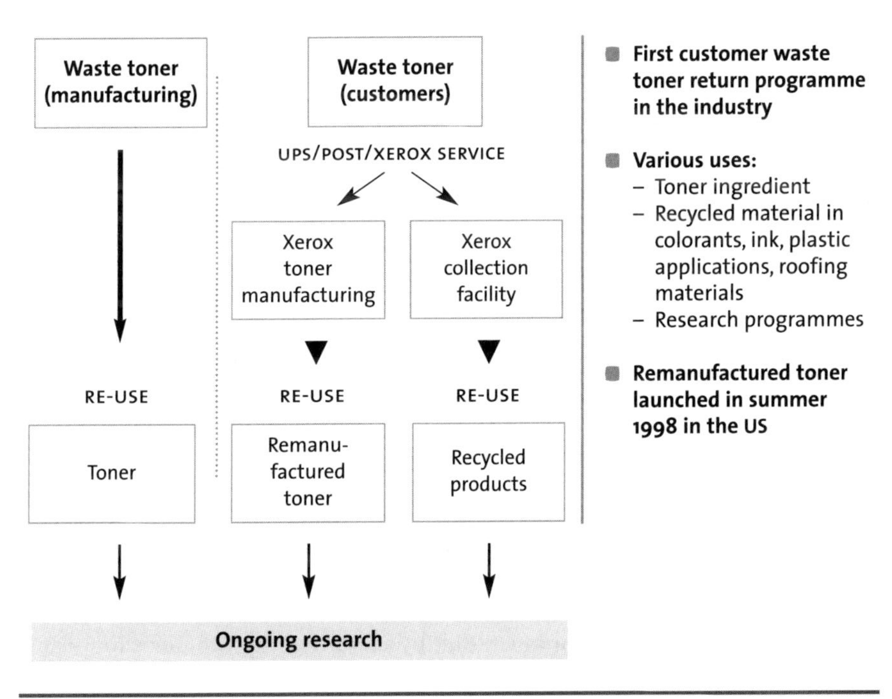

Figure 8 **Waste-free products: toner return**

Waste-free factories

Waste-free applies also to our factories, and we have five criteria:

- 90% reduction in solid waste to landfills (baseline 1990 level)
- 90% reduction in air emissions (baseline 1990)
- 90% reduction in hazardous waste (baseline 1990)
- Incorporate 25% post-consumer recycled materials in parts and packaging
- Energy usage within 10% of optimum level for facility

For landfill reductions, our performance is illustrated in Figure 9.

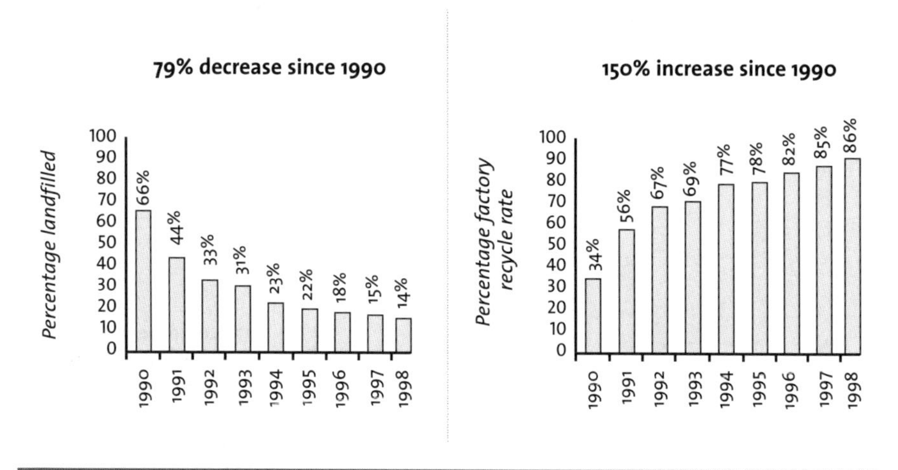

Figure 9 **Waste-free products: landfill reductions**

Partnerships with customers and suppliers

It has become clear to us now, after several years of thinking about how to integrate environmental practices into Xerox operations, that true sustainability depends on developing partnerships with our customers and our suppliers. Perhaps that is part of what leadership is all about. In any case, we have developed a stringent set of requirements for all of our suppliers worldwide, and we are actively involving our customers in sustainable operations. For us, for anyone moving in this direction, this is a whole new way of operating. It is not always easy to follow the trail of a product. But, by moving deep into our supply and customer chain, we are able to advance our ability to manage environmental impacts and move toward sustainability. This is not an easy challenge at all, particularly when we seek to maintain the same standards worldwide.

For example, the Xerox manufacturing facility in Mitcheldean, UK, held an environmental conference for suppliers in 1998 where we outlined the benefits of environmental leadership to 60 major suppliers of electronics, packaging, metal, plastics and rubber commodities.

We also developed 'EcoWorx', a document management software tool, which we have offered to customers as one of our new 'knowledge solutions' (Fig. 10). At our Webster, New York, toner and developer plant, EcoWorx has reduced the hours spent on EH&S documentation processes by 30%. Why shouldn't Xerox's environmental and document management expertise be available to our customers? We feel that reduction of paperwork and gradual elimination of it altogether is on the horizon. Dematerialisation is something we're just beginning to think about, so we'll have more to say about it in the future.

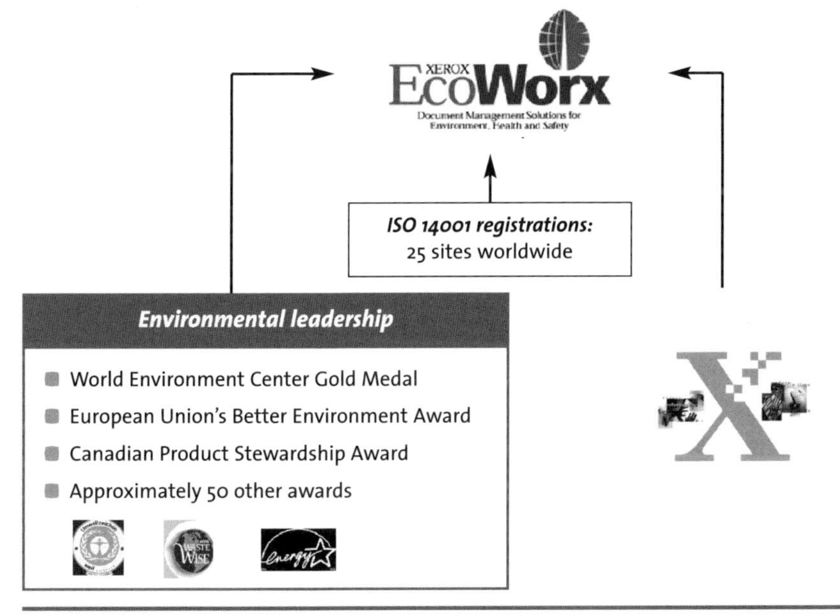

Figure 10 **'EcoWorx'**

Business results

Our asset recycling savings at Xerox have amounted to several hundred millions of dollars annually. Our 3R programme—reduce, re-use, recycle—produced $45 million in savings in 1998 alone, in addition to savings from equipment manufacture.

The evolution of Xerox's environmental leadership (Fig. 11) has taken us from emphasis on remediation and compliance to an eye on a sustainable future involving dematerialisation and knowledge management.

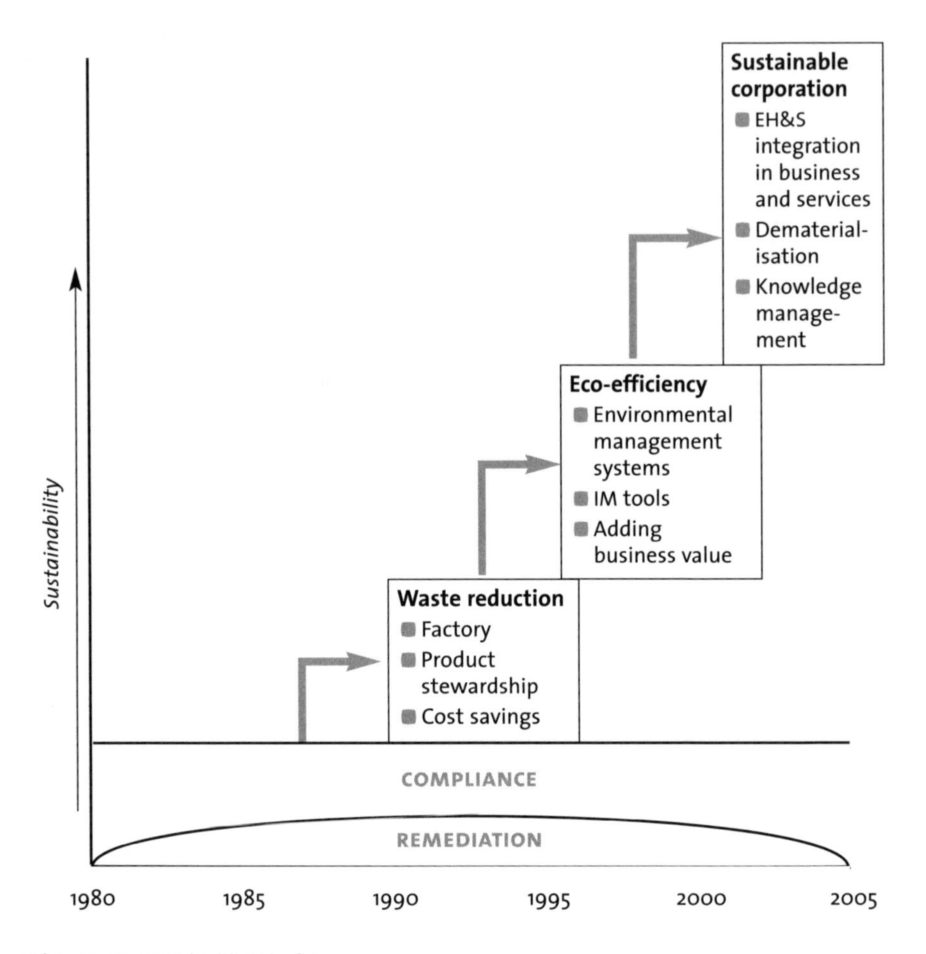

EH&S = environment, health and safety

Figure 11 **Evolution of environmental leadership**

ECO-EFFICIENCY
Harmonious co-existence

Canon Group's corporate philosophy, *kyosei* (harmonious co-existence), has translated into two decades of environmental innovations within the group, from improved resource productivity, to ecodesign, to green procurement (begun in 1997), to product recycling. Green procurement has become an important driver of environmental integration, and Canon has set the global standard in green procurement practices. Through such practices, Canon leverages environmental standards on more than 1,500 other companies that provide materials to Canon. Even though this supply-chain leveraging is a kind of partnership for change, we include Canon here under eco-efficiency because the partnership is within the industrial sector, not transversal in any sense (such as an industry–NGO partnership, for example).

Canon is now beginning to place labels on its products disclosing environmental impact information, a procedure known to have significantly altered the marketing and purchasing of appliances such as refrigerators and washing machines in the United States.

Yusuke Emura, Managing Director of Canon Group, joined Canon after completing his studies in production mechanical engineering at Kyu-shu University. He worked in Canon's mechanical department, then in copy machine engineering. He eventually moved into management positions until reaching his current position as Managing Director of Canon. He delivered the following presentation in Paris, in 1999.

ENVIRONMENTAL MANAGEMENT OF CANON GROUP

Yusuke Emura

Basic awareness of environmental issues

In the 20th century, the human race encountered three major problems of great difficulty. First is the East–West problem of ideological disparity, which we have seen resolved; second is the wealth–poverty gap between North and South, for which we have not yet found a solution; and the third involves the Earth's environment. While the first two problems should be resolved for the sake of people trying to co-exist in harmony and enjoy mutual prosperity with other people, the third should be resolved for the sake of people trying to co-exist in harmony with the Earth and nature. Unfortunately, the will of the Earth is not being reflected in our society. This is because, until recently, humans have not even been aware of the existence of anything other than themselves with which they should try to co-exist. The Earth does not openly make its own intentions known. Therefore, we who have an involvement with the environment need to be conscious of this and to express consideration toward the environment in our actions, in other words, to 'internalise' the environment.

Protecting the Earth's environment does not mean people protecting the Earth; it means changing ourselves so that we humans and our society can co-exist with it. The fact is that evidence of global warming, acid rain and pollution of underground water (Group A), along with the drying-up of underground resources, deforestation, species extinction (Group B), have already become visible around us. Each of these phenomena in the two groups has differing causes. Examples of pollution of the natural world have been placed in Group A while Group B includes phenomena resulting from the consumption of resources beyond the critical limit. Although caused by different factors, they have the same result: the exhaustion of resources. Atmospheric, water, soil and other types of pollution can only be called the worst possible form of waste. On the other hand, if we consider other aspects of the Earth's resources, we find the undeniable rule of providence that (1) all ecosystems are driven by the Earth's resources, and (2) other ecosystems are also resources for humans. These truths, through the emergence of environmental problems, amount to a warning: that we need to be aware that our Earth has a limited amount of resources for humanity in the present time and these resources must be shared among all ecosystems. The term 'internalising the environment' suggests that we should actualise this awareness in our personal values and lifestyles, and in the economic and other systems we form in our societies.

Environmental management of Canon Group

Corporate philosophy and environmental policy

The Canon Group's corporate philosophy is *kyosei*, or harmonious co-existence—the aim for harmony, co-existence and mutual prosperity with all things that affect oneself. Put another way, this philosophy itself actually means environmental protection of the Earth. The various business activities aimed at implementing this philosophy start with the five principles of research and development, which are the starting point of manufacturing products, and are carried out in accordance with a basic policy embodied in 'EQCD' (Fig. 12). One of the five principles of research and development states that Canon will not conduct research that goes against the environment. As for the basic policy, E stands for environment or ecology, Q for quality/performance, C for cost and D for delivery/supply.

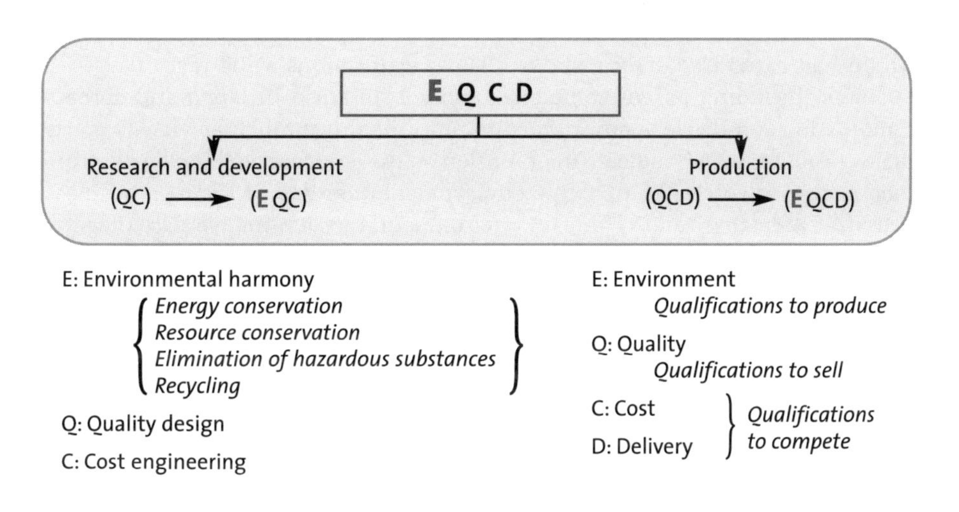

Figure 12 **Basic policy**

Traditional corporate policies have focused only on the elements affecting management efficiency—Q, C and D—in their business activities, and as a result have brought on the destruction of the environment we are seeing today. The EQCD policy means being aware that the pursuit of Q, C and D are allowable only under conditions that fulfil environmental protection and that businesses that cannot protect the environment have no right to manufacture products. Under the EQCD policy, the E must be given higher priority than management efficiency.

Improving resource productivity

Based on this philosophy and policy, the Canon Group is making efforts to co-exist in harmony with the environment. In these efforts, we are mindful that the essence of environmental problems lies in the lack of awareness of the limited nature of the Earth's resources and the waste of resources beyond the critical limit. We believe that improving resource productivity will counteract these negatives. We believe that basic measures are (1) energy saving, (2) resource saving and (3) elimination of hazardous substances. Our manufacturing based on this standpoint is aimed at improving resource productivity and is grounded in the process of reforming and innovating current systems from their very foundations. Our current development, production and sales systems which constitute our main functions as a product manufacturer are being rebuilt into new ecodesign, inverse manufacturing and eco-sales systems. These new systems will be supported by eco-research, eco-technology development and green procurement. Each system is maintained by two main activities, but each has the common objective of achieving energy saving, resource saving and elimination of hazardous substances (Fig. 13).

From the standpoint of environmental management, the building of an environmental management system (EMS), the disclosure of environmental information and environmental accounting are necessary. By the end of 1996 the Canon Group had earned ISO 14001 and/or EMAS certifications at its main places of business (including parent companies and subsidiaries) in Japan and abroad. Canon will soon disclose environmental information through eco-labels on its major products which indicate their burden on the environment and an environmental report which describes the environmental burden of Canon's business activities. A database and system for environmental accounting are also currently in the construction process.

Ecodesign

Products are the final fruits of business activities undertaken for the purpose of manufacturing products. The most important aspect of environmental protection involving all business activities is the reduction of the burden on the environment of the product itself. In other terms, this means the improvement of the product eco-efficiency. Practical effectiveness is improved through giving consideration to materials themselves (materials and parts) and carrying out environ-

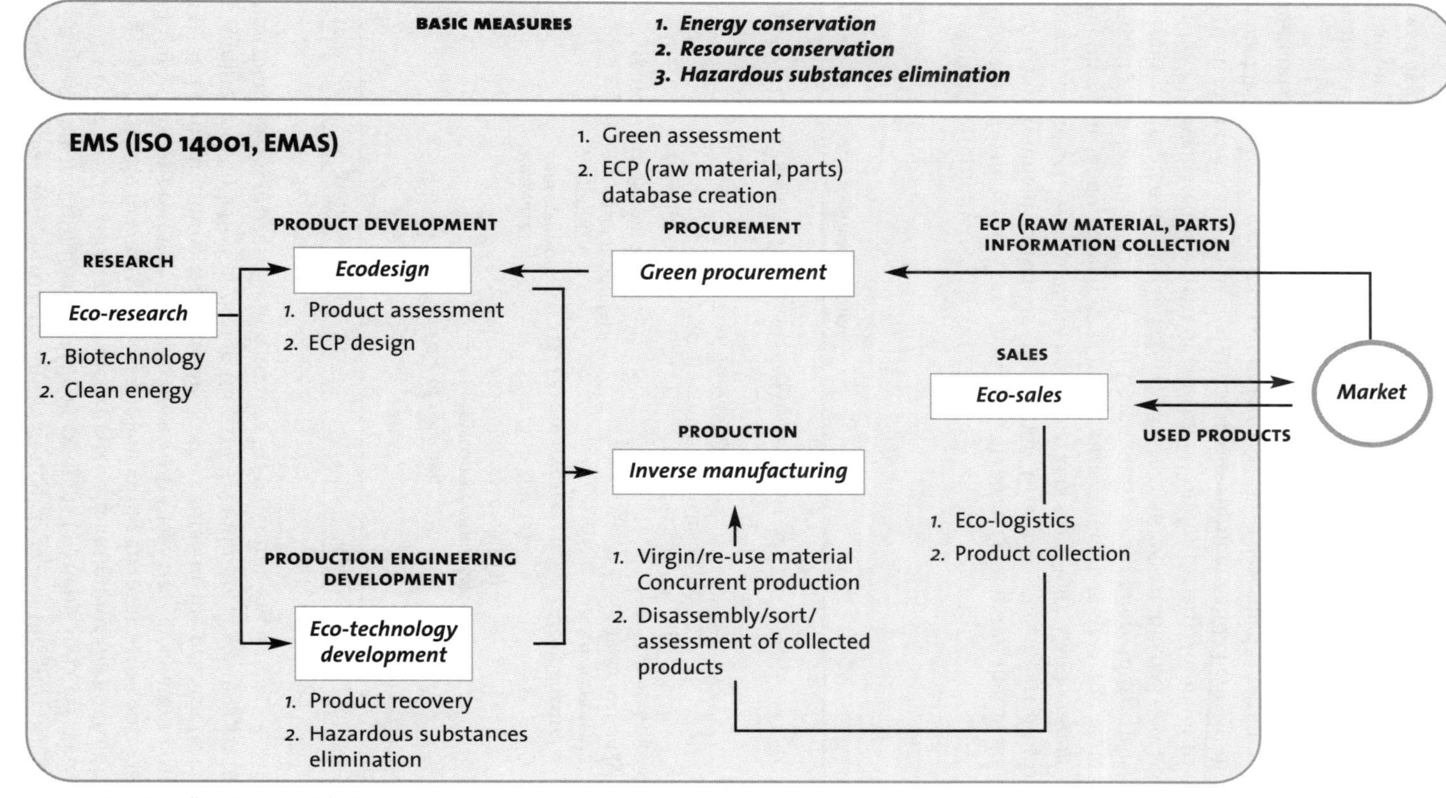

ECP = environmentally conscious products

Figure 13 **System of resource efficiency improvement**

ment-conscious design from the initial design stage. For this purpose, the Canon Group established product eco-efficiency items (55 items in 11 areas) as requirements to be met by products made with consideration for the environment (see Fig. 14). Since then, we have been implementing environment-conscious design of our group of products in accordance with these new evaluation criteria (Fig. 15), and applying three assessments in the course of the process from product planning to completion of design (Fig. 16). In addition, since 1997 we have been practising green procurement activities for obtaining materials.

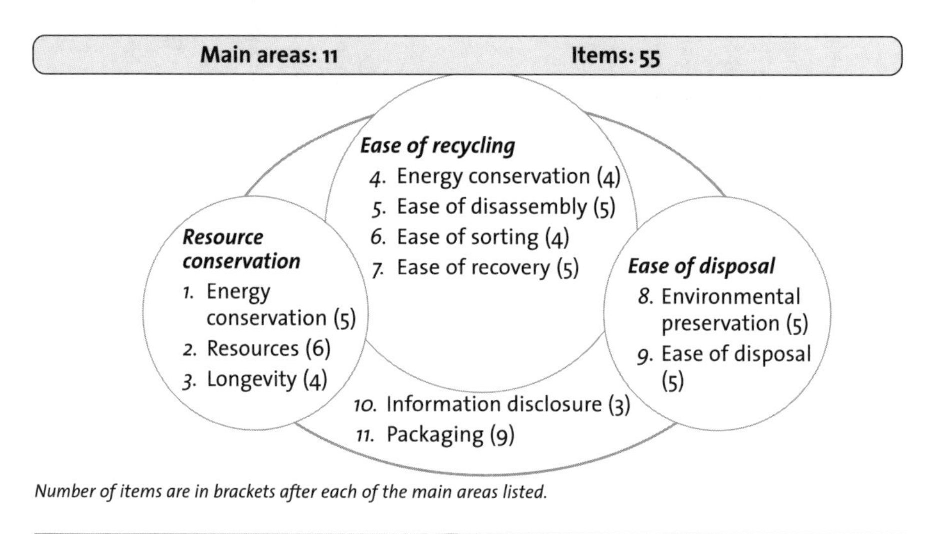

Number of items are in brackets after each of the main areas listed.

Figure 14 **Product eco-efficiency items**

Green procurement

To source the materials (raw materials, materials and parts) that make products more environment-conscious, the Canon Group has been carrying out green procurement activities in the world's markets since 1997. Canon requested 1,200 suppliers in Japan, which is a major procurement market for materials, 200 suppliers in South-East Asia and China and 100 suppliers in North America to supply environment-conscious products. In 1996, we established new assessment criteria to evaluate the extent of consideration for the environment in materials (green materials) and indicated these to our suppliers and clients in the form of a booklet, *Green Procurement Standards*. We also described concrete transaction methods in a *Green Procurement Standards Guidebook*.

The booklet of standards describes criteria for assessing the 'greenness' of a material. The criteria are comprised of seven basic parameters for corporate structure and 11 basic parameters for products themselves. The level of 'greenness' is determined from the sum of the product's score in each of these two criteria classifications. The parameters perceived as particularly important are:

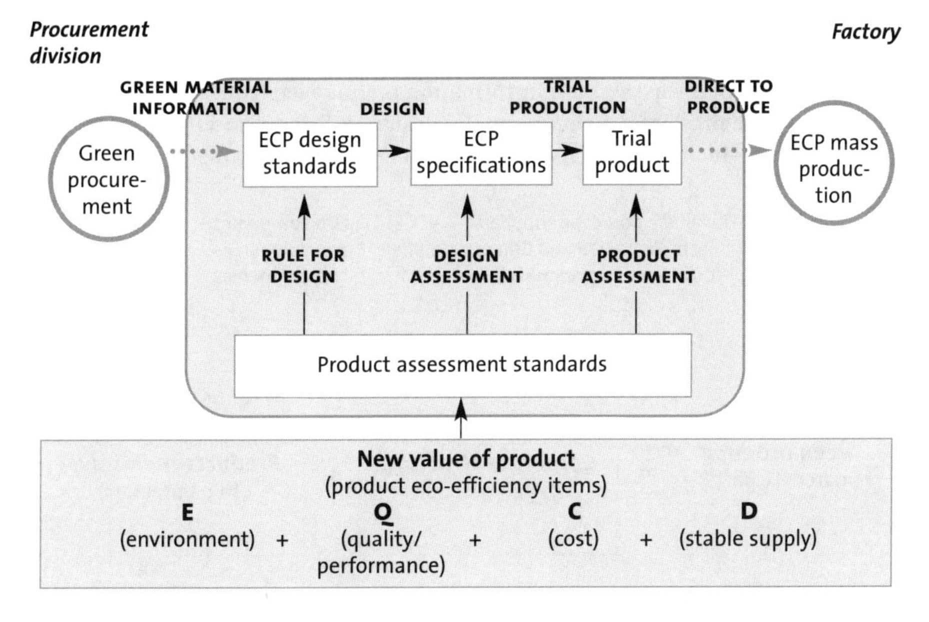

ECP = environmentally conscious products

Figure 15 **Ecodesign system**

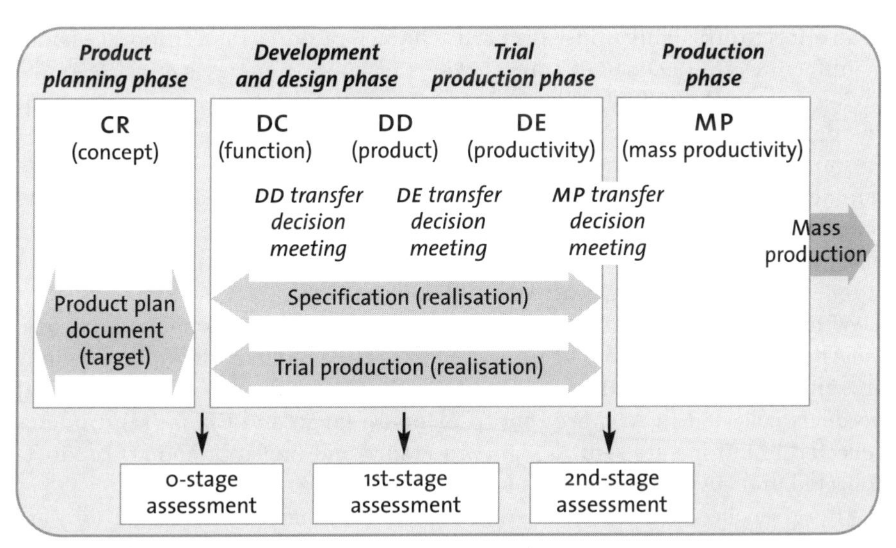

The three phases of product assessment apply to all products: (1) product planning; (2) development and design; and (3) trial production.

Figure 16 **Product assessment flow**

for corporate structure, the establishment or lack of an environment management system; and, for products themselves, the presence or lack of harmful chemical substances. All of the assessment information is conveyed from the procurement divisions to the product development and design divisions, and forms an important database which is useful for carrying out environment-conscious design.

Japanese domestic suppliers	1,200 companies
South-East Asian and Chinese suppliers	200 companies
North American suppliers	100 companies

PRODUCT EVALUATION CRITERIA

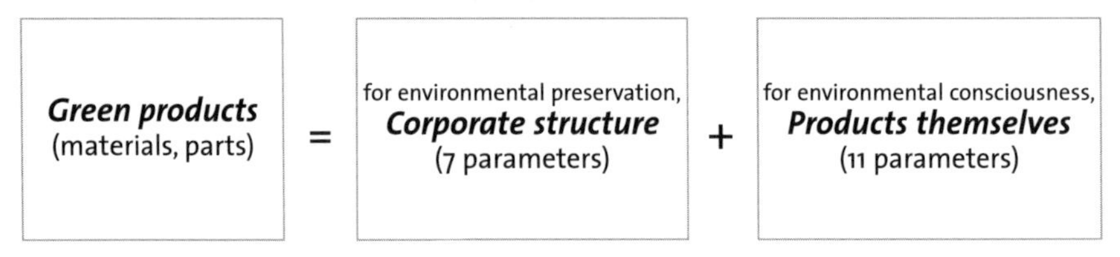

Note: The level of 'greenness' of products will be determined from the overall quantified score in individual evaluation parameters.

Figure 17 **Green procurement**

Inverse manufacturing (product recycling)

In an effort to efficiently utilise the Earth's limited resources as a corporation that manufactures products, the Canon Group is promoting energy saving, resource saving and the elimination of hazardous substances in the product development and design stages and is strongly promoting product recycling, which occupies an important position among activities to improve resource productivity. We initiated copying machine recycling programmes in major world markets in the 1980s, began recycling toner cartridges for copying machines, printers and fax machines in 1990, and started to recycle bubble-jet printer cartridges in 1996 (the latter as a pilot programme carried out in Japan).

We divide the world market for Canon products into four regions: Asia, Europe, the Americas and Oceania. The used products recovered in each of these regions are sent to recycling plants located in three areas under the following system: (1) products collected in Asia are sent to plants in Japan and China; (2) products collected in Europe are sent to plants in France and England; and (3) products collected in the Americas are sent to plants in North America (Fig. 18).

After the collected products are disassembled, cleaned and inspected, they are re-used as recycled materials and re-used parts which are utilised along with virgin materials and parts. Since collection systems differ by region, Canon has developed a unique integrated system which combines reverse distribution from sales companies, alliances with private transportation companies and the utilisa-

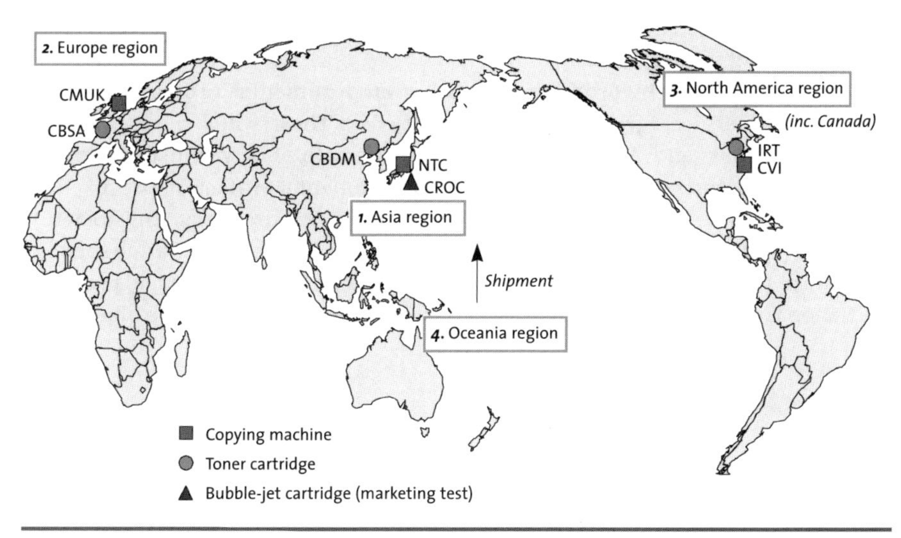

Figure 18 **Product recycling centres**

tion of postal systems. The amount collected also differs by product (Table 1 gives details relating to toner cartridges). The yearly average is from five to nine million units, and the percentage of these by weight which are re-used as materials and/or parts is between 70% and 99%. The leftover items are used to produce heat or are disposed by burial.

Based on:	***Number of parts***	***Weight (g)***
1. Re-used parts	63 items (54%)	379 (37%)
2. Recovered parts (parts ⇒ material ⇒ parts)	6 items (5%)	389 (38%)
3. Material recovery (parts ⇒ material)	25 items (22%)	245 (24%)
4. Disposal	22 items (19%)	9 (1%)

Cartridge (1 unit)
Total weight: 1,022 g; number of total parts: 116 items

Table 1 **Toner cartridge recycling (typical type)**

Conclusion

The focus of environmental protection by the Canon Group is shifting from individually attainable goals (such as the establishment of an environmental management system, the elimination of environment-polluting substances, the reduction of greenhouse effect gases and ECP design) to areas that require co-operation by society at large (such as green procurement, product recycling and the disclosure of environmental information).

Currently, our watchwords for promoting environmental protection are (1) recycling, (2) service, (3) co-operation and (4) risk. These watchwords speak of sharing awareness and strengthening co-operation with society, both of which are necessary for assuring the effectiveness of environmental protection practised by the Canon Group. This means sharing resources (a resource-recycling society), hardware (a society that rejects materialism), information (a co-operative society) and nature (a society that co-exists with nature). As a member of such a society, the Canon Group will strive to contribute to the achievement of these objectives.

ECO-EFFECTIVENESS

Eco-efficiency, now an accepted industrial solution, is not the promised panacea, say the authors of this next essay. They make an argument for 'eco-effectiveness' as a new metaphor. German-born chemist Michael Braungart was early on an environmental activist with Greenpeace International. Later, he invented a method of bleaching paper with oxygen instead of chlorine, thus eliminating the release of carcinogenic dioxins in the atmosphere. His method has been widely adopted in European and American paper production. Braungart was also involved in 1987 in developing the famous non-chlorofluorocarbon refrigerator.

William McDonough is an architect whose firm, William McDonough & Partners, uses innovative ecological design, such as interior trees, daylight illumination, roofs made of grass, and raised floors that allow for cool air to pass underneath. ('The Next Industrial Revolution' originally appeared in *The Atlantic Monthly*, October 1998, in a slightly longer version.)

THE NEXT INDUSTRIAL REVOLUTION

William McDonough and Michael Braungart

In the spring of 1912 one of the largest moving objects ever created by human beings left Southampton and began gliding toward New York. It was the epitome of its industrial age—a potent representation of technology, prosperity, luxury and progress. It weighed 66,000 tons. Its steel hull stretched the length of four city blocks. Each of its steam engines was the size of a townhouse. And it was headed for a disastrous encounter with the natural world. This vessel, of course, was the *Titanic*—a brute of a ship, seemingly impervious to the details of nature. In the minds of the captain, the crew and many of the passengers, nothing could sink it.

One might say that the infrastructure created by the Industrial Revolution of the 19th century resembles such a steamship. It is powered by fossil fuels, nuclear reactors and chemicals. It is pouring waste into the water and smoke into the sky. It is attempting to work by its own rules, contrary to those of the natural world. And, although it may seem invincible, its fundamental design flaws presage disaster. Yet many people still believe that, with a few minor alterations, this infrastructure can take us safely and prosperously into the future.

At the 1992 Earth Summit in Rio de Janeiro, many industrial participants touted a particular strategy: eco-efficiency. The machines of industry would be refitted with cleaner, faster, quieter engines. Prosperity would remain unobstructed and economic and organisational structures would remain intact. The hope was that eco-efficiency would transform human industry from a system that takes, makes and wastes into one that integrates economic, environmental and ethical concerns. Eco-efficiency is now considered by industries across the globe to be the strategy of choice for change.

What is eco-efficiency? Primarily, the term means 'doing more with less'—a precept that has its roots in early industrialisation. Henry Ford was adamant about lean and clean operating policies; he saved his company money by recycling and re-using materials, reduced the use of natural resources, minimised packaging, and set new standards with his time-saving assembly line. Ford wrote in 1926, 'You must get the most out of the power, out of the material, and out of the time'—a credo that could hang today on the wall of any eco-efficient factory.

The term 'eco-efficiency' was promoted by the Business Council (now the World Business Council) for Sustainable Development, a group of 48 industrial sponsors including Dow, DuPont, Con Agra and Chevron, who brought a business perspective to the Earth Summit. The council presented its call for change in practical terms, focusing on what businesses had to gain from a new ecological awareness rather than on what the environment had to lose if industry continued in current patterns. In *Changing Course*, a report released just before the summit, the group's founder, Stephan Schmidheiny, stressed the importance of eco-efficiency for all companies that aimed to be competitive, sustainable and successful over the long term (Schmidheiny 1992). In 1996 Schmidheiny said, 'I predict that within a decade it is going to be next to impossible for a business to be competitive without also being "eco-efficient".'

As Schmidheiny predicted, eco-efficiency has been working its way into industry with extraordinary success. The corporations committing themselves to it continue to increase in number, and include such big names as Monsanto, 3M and Johnson & Johnson. Its famous three Rs—reduce, re-use, recycle—are steadily gaining popularity in the home as well as the workplace. The trend stems in part from eco-efficiency's economic benefits, which can be considerable: 3M, for example, has saved more than $750 million through pollution-prevention projects, and other companies, too, claim to be realising big savings. Naturally, reducing resource consumption, energy use, emissions and wastes has implications for the environment as well. When one hears that DuPont has cut its emissions of airborne cancer-causing chemicals by almost 75% since 1987, one can't help feeling more secure. This is another benefit of eco-efficiency: it diminishes guilt and fear. By subscribing to eco-efficiency, people and industries can be less 'bad' and less fearful about the future. Or can they?

Eco-efficiency is an outwardly admirable and certainly well-intended concept, but, unfortunately, it is not a strategy for success over the long term, because it does not reach deep enough. It works within the same system that caused the problem in the first place, slowing it down with moral proscriptions and punitive demands. It presents little more than an illusion of change. Relying on eco-efficiency to save the environment will in fact achieve the opposite—it will let industry finish off everything quietly, persistently and completely.

We are forwarding a reshaping of human industry—the Next Industrial Revolution. Leaders of this movement include many people in diverse fields, among them commerce, politics, the humanities, science, engineering and education. As an architect and industrial designer, and a chemist, who have worked with both commercial and ecological systems, we see conflict between industry and the environment as a design problem—a very big design problem.

Many of the basic intentions behind the Industrial Revolution were good ones, which most of us would probably like to see carried out today: to bring more goods and services to larger numbers of people, to raise standards of living, and to give people more choice and opportunity, among others. But there were crucial omissions. Perpetuating the diversity and vitality of forests, rivers, oceans, air, soil and animals was not part of the agenda.

If someone were to present the Industrial Revolution as a retroactive design assignment, it might sound like this:

Design a system of production that

- Puts billions of pounds of toxic material into the air, water and soil
- Measures prosperity by activity, not legacy
- Requires thousands of complex regulations to keep people and natural systems from being poisoned too quickly
- Produces materials so dangerous that they will require constant vigilance from future generations
- Results in gigantic amounts of waste
- Puts valuable materials in holes all over the planet, where they can never be retrieved
- Erodes the diversity of biological species and cultural practices

Eco-efficiency instead

- Releases *fewer* pounds of toxic material into the air, water, and soil every year
- Measures prosperity by *less* activity
- *Meets or exceeds* the stipulations of thousands of complex regulations that aim to keep people and natural systems from being poisoned too quickly
- Produces *fewer* dangerous materials that will require constant vigilance from future generations
- Results in *smaller* amounts of waste
- Puts *fewer* valuable materials in holes all over the planet, where they can never be retrieved
- Standardises and homogenises biological species and cultural practices

Plainly put, eco-efficiency aspires to make the old, destructive system less so. But its goals, however admirable, are fatally limited.

Reduction, re-use and recycling slow down the rates of contamination and depletion but do not stop these processes. Much recycling, for instance, is what we call 'downcycling', because it reduces the quality of a material over time. When plastic—other than that found in such products as soda and water bottles—is recycled, it is often mixed with different plastics to produce a hybrid of lower quality, which is then moulded into something amorphous and cheap, such as park benches or speed bumps. The original high-quality material is not retrieved, and it eventually ends up in landfills or incinerators.

The well-intended, creative use of recycled materials for new products can be misguided. For example, people may feel that they are making an ecologically sound choice by buying and wearing clothing made of fibres from recycled plastic bottles. But the fibres from plastic bottles were not specifically designed to be next to human skin. Blindly adopting superficial 'environmental' approaches without fully understanding their effects can be no better than doing nothing.

Recycling is more expensive for communities than it needs to be, partly because traditional recycling tries to force materials into more lifetimes than they were designed for—a complicated and messy conversion, and one that itself expends energy and resources. Very few objects of modern consumption were designed with recycling in mind. If the process is truly to save money and materials, products must be designed from the very beginning to be recycled or even 'upcycled'—a term we use to describe the return to industrial systems of materials with improved, rather than degraded, quality.

The reduction of potentially harmful emissions and wastes is another goal of eco-efficiency. But current studies are beginning to raise concern that even tiny amounts of dangerous emissions can have disastrous effects on biological systems over time. This is a particular concern in the case of endocrine disrupters—industrial chemicals in a variety of modern plastics and consumer goods which appear to mimic hormones and connect with receptors in human beings and other organisms. Theo Colborn, Dianne Dumanoski and John Peterson Myers, the authors of *Our Stolen Future* (1996), a ground-breaking study on certain synthetic chemicals and the environment, assert that 'astoundingly small quantities of these hormonally active compounds can wreak all manner of biological havoc, particularly in those exposed in the womb'.

On another front, new research on particulates—microscopic particles released during incineration and combustion processes, such as those in power plants and automobiles—shows that they can lodge in and damage the lungs, especially in children and the elderly. A 1995 Harvard study found that as many as 100,000 people die annually as a result of these tiny particles. Although regulations for smaller particles are in place, implementation does not have to begin until 2005. Real change would be not regulating the release of particles but attempting to eliminate dangerous emissions altogether—by design.

Applying nature's cycles to industry

'Produce more with less', 'minimise waste', 'reduce' and similar dictates advance the notion of a world of limits—one whose carrying capacity is strained by burgeoning populations and exploding production and consumption. Eco-efficiency tells us to restrict industry and curtail growth—to try to limit the creativity and productiveness of humankind. But the idea that the natural world is inevitably destroyed by human industry, or that excessive demand for goods and services causes environmental ills, is a simplification. Nature—highly industrious, astonishingly productive and creative, even 'wasteful'—is not efficient but *effective.*

Consider the cherry tree. It makes thousands of blossoms just so that another tree might germinate, take root and grow. Who would notice piles of cherry blossoms littering the ground in the spring and think, 'How inefficient and wasteful'? The tree's abundance is useful and safe. After falling to the ground, the blossoms return to the soil and become nutrients for the surrounding environment. Every last particle contributes in some way to the health of a thriving ecosystem. 'Waste equals food'—the first principle of the Next Industrial Revolution.

The cherry tree is just one example of nature's industry, which operates according to cycles of nutrients and metabolisms. This cyclical system is powered by the sun and constantly adapts to local circumstances. Waste that stays waste does not exist.

Human industry, on the other hand, is severely limited. It follows a one-way, linear, cradle-to-grave manufacturing line in which things are created and eventually discarded, usually in an incinerator or a landfill. Unlike the waste from nature's work, the waste from human industry is not 'food' at all. In fact, it is often poison. Thus the two conflicting systems: a pile of cherry blossoms and a heap of toxic junk in a landfill.

But there is an alternative—one that will allow both business and nature to be fecund and productive. This alternative is what we call 'eco-effectiveness'. Our concept of eco-effectiveness leads to human industry that is regenerative rather than depletive. It involves the design of things that celebrate interdependence with other living systems. From an industrial-design perspective, it means products that work within cradle-to-cradle life-cycles rather than cradle-to-grave ones.

Waste equals food

Ancient nomadic cultures tended to leave organic wastes behind, restoring nutrients to the soil and the surrounding environment. Modern, settled societies simply want to get rid of waste as quickly as possible. The potential nutrients in organic waste are lost when they are disposed of in landfills, where they cannot be used to rebuild soil; depositing synthetic materials and chemicals in natural systems strains the environment. The ability of complex, interdependent natural ecosystems to absorb such foreign material is limited if not non-existent. Nature cannot do anything with the stuff *by design*: many manufactured products are intended not to break down under natural conditions.

If people are to prosper within the natural world, all the products and materials manufactured by industry must, after each useful life, provide nourishment for something new. Since many of the things people make are not natural, they are not safe 'food' for biological systems. Products composed of materials that do not biodegrade should be designed as technical nutrients that continually circulate within closed-loop industrial cycles—the technical metabolism.

In order for these two metabolisms to remain healthy, great care must be taken to avoid cross-contamination. Things that go into the biological metabolism should not contain mutagens, carcinogens, heavy metals, endocrine disrupters, persistent toxic substances or bioaccumulative substances. Things that go into the technical metabolism should be kept well apart from the biological metabolism.

If the things people make are to be safely channelled into one or the other of these metabolisms, then products can be considered to contain two kinds of material: *biological nutrients* and *technical nutrients.*

Biological nutrients will be designed to return to the organic cycle—to be literally consumed by microorganisms and other creatures in the soil. Most packaging (which makes up about 50% by volume of the solid-waste stream) should be composed of biological nutrients—materials that can be tossed onto the ground or the compost heap to biodegrade. There is no need for shampoo bottles, toothpaste tubes, yoghurt cartons, juice containers and other packaging to last decades (or even centuries) longer than what came inside them.

Technical nutrients will be designed to go back into the technical cycle. Right now anyone can dump an old television into a trash can. But the average television is made of hundreds of chemicals, some of which are toxic. Others are valuable nutrients for industry, which are wasted when the television ends up in a landfill. The re-use of technical nutrients in closed-loop industrial cycles is distinct from traditional recycling, because it allows materials to retain their quality: high-quality plastic computer cases would continually circulate as high-quality computer cases, instead of being downcycled to make soundproof barriers or flowerpots.

Customers would buy the *service* of such products, and when they had finished with the products, or simply wanted to upgrade to a newer version, the manufacturer would take back the old ones, break them down, and use their complex materials in new products.

First fruits: a biological nutrient

A few years ago we helped to conceive and create a compostable upholstery fabric—a biological nutrient. We were initially asked by DesignTex to create an aesthetically unique fabric that was also ecologically intelligent—although the client did not quite know at that point what this would mean. The challenge helped to clarify, both for us and for the company we were working with, the difference between superficial responses such as recycling and reduction and the more significant changes required by the Next Industrial Revolution.

For example, when the company first sought to meet our desire for an environmentally safe fabric, it presented what it thought was a wholesome option: cotton, which is natural, combined with PET (polyethylene terephthalate) fibres from recycled beverage bottles. Since the proposed hybrid could be described with two important eco-buzzwords, 'natural' and 'recycled', it appeared to be environmentally ideal. The materials were readily available, market-tested, durable and cheap. But, when the project team looked carefully at what the manifestations of such a hybrid might be in the long run, we discovered some disturbing facts. When a person sits in an office chair and shifts around, the fabric beneath them abrades;

tiny particles of it are inhaled or swallowed by the user and other people nearby. PET was not designed to be inhaled. Furthermore, PET would prevent the proposed hybrid from going back into the soil safely, and the cotton would prevent it from re-entering an industrial cycle. The hybrid would still add junk to landfills, and it might also be dangerous.

The team decided to design a fabric so safe that one could literally eat it. The European textile mill chosen to produce the fabric was quite 'clean' environmentally, and yet it had an interesting problem: although the mill's director had been diligent about reducing levels of dangerous emissions, government regulators had recently defined the trimmings of his fabric as hazardous waste. We sought a different end for our trimmings: mulch for the local garden club. When removed from the frame after the chair's useful life and tossed onto the ground to mingle with sun, water, and hungry microorganisms, both the fabric and its trimmings would decompose naturally.

The team decided on a mixture of safe, pesticide-free plant and animal fibres for the fabric (ramie and wool) and began working on perhaps the most difficult aspect: the finishes, dyes and other processing chemicals. If the fabric was to go back into the soil safely, it had to be free of mutagens, carcinogens, heavy metals, endocrine disrupters, persistent toxic substances and bioaccumulative substances. Sixty chemical companies were approached about joining the project, and all declined, uncomfortable with the idea of exposing their chemistry to the kind of scrutiny necessary. Finally, one European company, Ciba-Geigy, agreed to join.

With that company's help the project team considered more than 8,000 chemicals used in the textile industry and eliminated 7,962. The fabric—in fact, an entire line of fabrics—was created using only 38 chemicals.

The director of the mill told a surprising story after the fabrics were in production. When regulators came by to test the effluent, they thought their instruments were broken. After testing the influent as well, they realised that the equipment was fine—the water coming out of the factory was as clean as the water going in. The manufacturing process itself was filtering the water. The new design not only bypassed the traditional 3R responses to environmental problems but also eliminated the need for regulation.

In our Next Industrial Revolution, regulations can be seen as signals of design failure. They burden industry, by involving government in commerce and by interfering with the marketplace. Manufacturers in countries that are less hindered by regulations, and whose factories emit *more* toxic substances, have an economic advantage: they can produce and sell things for less. If a factory is not emitting dangerous substances and needs no regulation, and can thus compete directly with unregulated factories in other countries, that is good news environmentally, ethically and economically.

A technical nutrient

Someone who has finished with a traditional carpet must pay to have it removed. The energy, effort and materials that went into it are lost to the manufacturer; the carpet becomes little more than a heap of potentially hazardous petrochemicals that must be toted to a landfill. Meanwhile, raw materials must continually be extracted to make new carpets.

The typical carpet consists of nylon embedded in fibreglass and PVC (polyvinyl chloride) After its useful life a manufacturer can only downcycle it—shave off some of the nylon for further use and melt the leftovers. The world's largest commercial carpet company, Interface, is adopting our technical-nutrient concept with a carpet designed for complete recycling. When a customer wants to replace it, the manufacturer simply takes back the technical nutrient—depending on the product, either part or all of the carpet—and returns a carpet in the customer's desired colour, style and texture. The carpet company continues to own the material but leases it and maintains it, providing customers with the *service* of the carpet. Eventually the carpet will wear out like any other, and the manufacturer will re-use its materials at their original level of quality or a higher one.

The advantages of such a system, widely applied to many industrial products, are twofold: no useless and potentially dangerous waste is generated, as it might still be in eco-efficient systems, and billions of dollars' worth of valuable materials are saved and retained by the manufacturer.

Selling intelligence, not poison

Currently, chemical companies warn farmers to be careful with pesticides, and yet the companies benefit when more pesticides are sold. In other words, the companies are unintentionally invested in wastefulness and even in the mishandling of their products, which can result in contamination of the soil, water and air. Imagine what would happen if a chemical company sold intelligence instead of pesticides—that is, if farmers or agro-businesses paid pesticide manufacturers to protect their crops against loss from pests instead of buying dangerous regulated chemicals to use at their own discretion. It would in effect be buying crop insurance. Farmers would be saying, 'I'll pay you to deal with boll weevils, and you do it as intelligently as you can.' At the same price per acre, everyone would still profit. The pesticide purveyor would be invested in *not* using pesticide, to avoid wasting materials. Furthermore, since the manufacturer would bear responsibility for the hazardous materials, it would have incentives to come up with less dangerous ways to get rid of pests. Farmers are not interested in handling dangerous chemicals; they want to grow crops. Chemical companies do not want to contaminate soil, water and air; they want to make money.

Consider the unintended design legacy of the average shoe. With each step of your shoe the sole releases tiny particles of potentially harmful substances that may contaminate and reduce the vitality of the soil. With the next rain these particles will wash into the plants and soil along the road, adding another burden to the environment.

Shoes could be redesigned so that the sole was a biological nutrient. When it broke down under a pounding foot and interacted with nature, it would nourish the biological metabolism instead of poisoning it. Other parts of the shoe might be designed as technical nutrients, to be returned to industrial cycles. Most shoes—in fact, most products of the current industrial system—are fairly primitive in their relationship to the natural world. With the scientific and technical tools currently available, this need not be the case.

Respect diversity and use the sun

A leading goal of design in this century has been to achieve universally applicable solutions. In the field of architecture the International Style is a good example. As a result of the widespread adoption of the International Style, architecture has become uniform in many settings. That is, an office building can look and work the same anywhere. Materials such as steel, cement and glass can be transported all over the world, eliminating dependence on a region's particular energy and material flows. With more energy forced into the heating and cooling system, the same building can operate similarly in vastly different settings.

The second principle of the Next Industrial Revolution is 'respect diversity'. Designs will respect the regional, cultural and material uniqueness of a place. Wastes and emissions will regenerate rather than deplete, and design will be flexible, to allow for changes in the needs of people and communities. For example, office buildings will be convertible into apartments, instead of ending up as rubble in a construction landfill when the market changes.

The third principle of the Next Industrial Revolution is 'use solar energy'. Human systems now rely on fossil fuels and petrochemicals, and on incineration processes that often have destructive side-effects. Today even the most advanced building or factory in the world is still a kind of steamship, polluting, contaminating and depleting the surrounding environment, and relying on scarce amounts of natural light and fresh air. People are essentially working in the dark, and they are often breathing unhealthful air. Imagine, instead, a building as a kind of tree. It would purify air, accrue solar income, produce more energy than it consumes, create shade and habitat, enrich soil, and change with the seasons. Oberlin College is currently working on a building that is a good start: it is designed to make more energy than it needs to operate and to purify its own waste-water.

Equity, economy, ecology

The Next Industrial Revolution incorporates positive intentions across a wide spectrum of human concerns. People within the sustainability movement have found that three categories are helpful in articulating these concerns: equity, economy and ecology.

Equity refers to social justice. Does a design depreciate or enrich people and communities? Shoe companies have been blamed for exposing workers in factories overseas to chemicals in amounts that exceed safe limits. Eco-efficiency would reduce those amounts to meet certain standards; eco-effectiveness would not use a potentially dangerous chemical in the first place. What an advance for humankind it would be if no factory worker anywhere worked in dangerous or inhumane conditions.

Economy refers to market viability. Does a product reflect the needs of producers and consumers for affordable products? Safe, intelligent designs should be affordable by and accessible to a wide range of customers, and profitable to the company that makes them, because commerce is the engine of change.

Ecology, of course, refers to environmental intelligence. Is a material a biological nutrient or a technical nutrient? Does it meet nature's design criteria: waste equals food, respect diversity and use solar energy?

The Next Industrial Revolution can be framed as the following assignment: design an industrial system for the next century that

- Introduces no hazardous materials into the air, water or soil
- Measures prosperity by how much natural capital we can accrue in productive ways
- Measures productivity by how many people are gainfully and meaningfully employed
- Measures progress by how many buildings have no smokestacks or dangerous effluents
- Does not require regulations whose purpose is to stop us from killing ourselves too quickly
- Produces nothing that will require future generations to maintain vigilance
- Celebrates the abundance of biological and cultural diversity and solar income

Albert Einstein wrote, 'The world will not evolve past its current state of crisis by using the same thinking that created the situation.' Many people believe that new industrial revolutions are already taking place, with the rise of cybertechnology, biotechnology and nanotechnology. It is true that these are powerful tools for change. But they are only tools—hyperefficient engines for the steamship of the first Industrial Revolution. Similarly, eco-efficiency is a valuable and laudable tool, and a prelude to what should come next. But it, too, fails to move us beyond the first revolution. It is time for designs that are creative, abundant, prosperous and intelligent from the start. The model for the Next Industrial Revolution may well have been right in front of us the whole time: a tree.

FROM PRODUCTS TO SERVICES

Another metaphor for change is the move 'from products to services', where the challenge is to disconnect economic growth from increased consumption of natural resources, by focusing on providing the service functions of products rather than on selling the products themselves. One such example of a business moving from products to services is the US carpet manufacturer Interface Inc., chaired by Ray Anderson, who explains in his book *Mid-Course Correction* (1998) how his company is becoming a 'sustainable enterprise'.

Following are excerpts from a talk given in Paris in 1997 by Ezio Manzini (Politecnico de Milano) on strategic design for a new product–service mix—with examples from businesses such as Rank-Xerox, Ciba-Geigy, Dow and Black & Decker. Environmental design, according to Ezio Manzini, is not just drawing the skeleton and outline of a product. Rather, it involves understanding the function of a product or service and then reinventing it with new rules, all of which require changes in the business-as-usual attitude. It is on this product–service cusp where much new action on sustainable development is taking place. For the industrial designer, environmental questions take their place alongside economic and aesthetic demands.

LEAPFROG
Short-term strategies for sustainability

Ezio Manzini

Sustainability, in its strongest sense, implies such a deep transformation of production and consumption activities as to represent a 'systemic discontinuity': a change that requires much more than the incremental innovation of technologies in use, or redesigning what exists. Sustainability is more than a partial modification of the existing non-sustainable ways in which industrial societies produce and consume. Therefore, achieving sustainability has to be considered a transition towards a new system of production and consumption requiring a new economy and a new culture.

From this perspective, 'short-term strategies for sustainability' means a set of initiatives whose goal is to favour transition as well as to make it take place in the most acceptable way for society, that is to say assuming both production continuity and social democracy.

Short-term strategies for sustainability

An environmental and industrial policy that leads to an increase by one point in the efficiency of existing car engines may be very significant in terms of immediate quantitative results, and therefore a very positive environmental policy. Nonetheless, it cannot be considered a strategy for sustainability. On the quantitative side, the increased environmental performance of existing engines, even if driven to their utmost possibility, does not get to the roots of the environmental and social problem of mobility. On the cultural side, to promote better efficiency of engines doesn't change the user's attitude towards mobility, nor the corporate culture of the manufacturer. For car industries, improving the efficiency of engines is the most normal and institutional activity that their research centres can do. Thus this policy, being entirely in keeping with traditional behaviours and business-as-usual routines, doesn't represent any discontinuity towards sustainability.

Let's now consider a policy that seeks to increase some form of sharing in car utilisation: from the idea of the collective taxi, to the promotion of car sharing or car pooling systems. Even if the overall quantitative results attainable by following this road would be almost negligible, a policy working in this direction is absolutely in line with the prospect of sustainability, and is to be seen as a short-term strategy to sustainability. Indeed, while for the time being involving only a minority of users, the shared use of cars presents a relevant point of cultural and behavioural discontinuity. Offering a relation with cars different from the one that has ruled to date is a potential contribution to defining a new idea of mobility.

Furthermore, in terms of an increased eco-efficiency per unit of service rendered, with the simple passage from one passenger per vehicle to two passengers per vehicle, a far better result is achieved than that which corps of engineers could have accomplished by working for years on the technology of engines.

Short-term strategies as 'leapfrog strategies'

Short-term strategies for sustainability are a set of initiatives promoting a direct shift from a non-sustainable mix of product and services to a sustainable one. This definition requires some explanation. In particular some implicit assumptions have to be made explicit:

- What users demand is not products or services, but the results that these products and services permit to be achieved.
- Given a result, it can be achieved by different combinations of products and services.
- A shift in consumption and behaviour patterns takes place if new and more sustainable combinations of products and services are recognised by users as better answers than existing ones, or because they meet a previously unanswered demand.

Leapfrog strategies in practice

In order to discuss the characteristics and the potential of short-term strategies for sustainability, some examples of their implementation follow. Examples will be grouped on the basis of two approaches: the result-oriented approach and the utility-oriented approach.

The result-oriented approach

Within this approach, the objective of a business becomes the sale of results rather than physical products. In this framework, which is also defined as the internalisation of the product, it is the producer's economic interest that pushes towards an increased eco-efficiency with the result of extending the life of products, components and materials, and optimising their utilisation.

The discontinuity that these initiatives put into effect lies mainly on the producer side. In fact, the shift from selling products to offering a mix of products and services asks for a deep change in a company's mission and organisation.

In the thermal management of buildings there now exist forms of contract (demand-side management and least-cost planning) whereby what is offered and guaranteed is a thermal comfort service and not the actual quantity of heating fuel supplied. What is interesting is that in all these cases the economic interest of the producer becomes that of guaranteeing the best possible service by reducing his own costs: that is, by reducing the consumption of fuel.

In the USA, for example, the energy service companies (ESCOs) supply an overall package of services on a turnkey basis. The package comprises: supply of energy resources, identification and selection of conservation measures, and installation, operation and maintenance of the energy supply. Something very similar is happening in Europe: RMM Energy GmbH is a company that has completed some 200 projects in Germany and in Switzerland since 1987. The Stadtwerke of the City of Hannover is performing one of the most comprehensive least-cost planning (LCP) projects financed by the European Commission (EC) in Europe and is co-ordinating ten LCP pilot programmes of other German utilities initiated by the SAVE Programme (a multi-annual programme for the promotion of energy efficiency) of the EC. In the Netherlands, the Energy Contract Partners are proposing the same range of services offered by the ESCOs in the USA.

A similar approach has been followed in other fields, such as agriculture, industrial painting and engineered materials.

In agriculture, a pest-control service can be offered instead of anti-pest chemicals. For example, Zeneca Group plc has developed an integrated pest management programme that can be tailored to the local situation. Its features are to correctly identify pests and natural enemies and to teach farmers when it is absolutely necessary to spray and how to promote non-chemical methods of control.

In the field of industrial paintings, Ciba is moving to become a supplier of colour services to its customers rather than merely selling dyes and pigments. In the field of engineered materials, the Cookson Group plc's central strategy is to retain the ownership of its product throughout its life-cycle. Alpha-Fry, a Cookson subsidiary in Germany, developed a take-back system for their solder paste packaging and a new kind of packaging. This new type of jar is of pure tin which, on return, is thrown into a melting pot and used as raw material in the manufacturing of solder bars.

The same attitude to selling results rather than products can be found in the field of durable goods and appliances. The most well-known case is that of Rank-Xerox.

Rank-Xerox GmbH proposed the introduction of a function compound consisting of the copier, its maintenance and repairs, the making of the copies and the collection and delivery of the original documents. This concept, realised for the first time in the head office of the German Henkel KGaA, is now a fundamental strategy of Rank-Xerox.

If the case of Rank-Xerox is the most clear, it is not the only one. An Italian company, Bibo, whose initial business was producing mono-use plastic dishes, has, since 1993, been providing a service for collective and public restoration services: they supply, take back and recycle plastic dishes, retaining the ownership of their products through the life-cycle.

The result-oriented approach can be extended also to cases in which the offered result is related only to some aspects of the producer–user relationship, as, for instance, the packaging. These cases are significant because, within this approach, some companies significantly change the way in which they operate and the interface that they have with their customers.

For instance, SafeChem is a new company established by Dow to deal with chlorinated solvents. Its activity includes product stewardship, and responsible care and distribution consulting services on how to change equipment to respond to the regulation in Germany and to recycle the used solvent. SafeChem has developed a new type of container, the Safetainer, to answer the demand of an effective closed-loop system. These are special containers with an air-tight pumping system which ensure that zero emissions occur during the transport phase and at the point when solvents are transferred from the containers into the distillation equipment. So they now sell de-greasing services rather than solvents. This shifts the incentive from selling more solvents to providing the least (and least toxic) solvents within a broader (and more value-added) package of de-greasing services.

The utility-oriented approach

Within this approach, the objective of a business becomes the promotion of leasing, pooling or sharing of certain goods, with the environmental advantage of optimising their use and reducing the quantity of products required to meet a given result. The leap that this approach puts into effect lies mainly on the user side and is the required change in the behaviour and consumption pattern (the shift from individual use and ownership of products to the leasing, pooling and sharing of services). Examples of new leasing services (do-it-yourself tools, sports equipment, computer and electronic devices) are an improvement of what traditionally exists.

Black & Decker GmbH offers an eco-leasing service for a professional series of mechanical saws, for the private handyman or even some professionals.

Car-sharing initiatives have a long story and have to be considered as forms of entrepreneurial activities. Car sharing can be related also to car-free cities projects. The development of these projects is in fact a significant activity that has to be seen as the realisation of a leapfrog strategy: a direct leap to the idea that, for some urban developments and for some specific target groups, being car-free could be an added value.

In Europe, for instance, there are several projects of car-free cities. For instance, Hollerland in Breme includes about 300 buildings and foresees some parking places for car sharing. Westerpark in Amsterdam and Stadthaus Schlump in Berlin are other examples and there are others in Monaco and Vienna.

FROM END-OF-PIPE TO INTEGRATION

Moving from end-of-pipe thinking to thinking about the integration of environmental concerns at the moment of conception of a project or product is an interesting historical change.

In the following excerpts from a 1997 talk in Paris, when he was German Minister of Towns, Planning and New Construction, Dr Klaus Töpfer presents himself as having moved from being an 'end-of-pipe' Minister of the Environment to a more 'integrated' Minister of Urban Affairs involved in the creation of the new Berlin. Dr Töpfer's career has given him a perspective that has evolved along with the issues and metaphors considered in these pages. He moved from German Minister of Environment to Director of the United Nations Commission on Sustainable Development, and then to Minister of Towns, Planning and New Construction. Now, Dr Töpfer has become the Executive Director of the United Nations Environment Programme (UNEP).

The second set of excerpts below is from a 1999 presentation in Paris by Dr Töpfer. Appropriately, now that his focus has become global, the metaphor that dominates in this second part is 'poverty: the most toxic environmental waste'.

REMARKS BY KLAUS TÖPFER

1997, 1999

1997

I became Environment Minister in Germany in 1987; before that I was Environment Minister for two years in one of our states, the Länder Rheinlandpfalz. At that time there was no risk at all in organising a conference on the environment; unlike today, when it is clearly not an easy task. Back then we automatically generated huge audiences talking about very simple topics such as waste management, air pollution, water and so on. So, looking back, I think things have changed a little, especially in Europe and perhaps in other parts of the world, because this is unfortunately no longer the case.

There is of course one explanation: that we succeeded in achieving our aims. I was at the Rio Summit of 1992 as head of the German delegation, and subsequently I was the second chairman of the Commission for Sustainable Development, the follow-up institution to the Rio Summit, at the United Nations in New York. Knowing what has happened since the Rio Summit, I am not so sure that we have achieved all that we wanted to, and I don't believe, by any means, that the fact we have accomplished our goals is the reason there are so few conferences.

On the other hand, I have no doubt that we have been facing up to many new challenges since the Rio Summit: for example, the historical development of the disappearance of a bipolar world; there is now no Soviet Union. Hearing this prediction in 1987, when I was appointed Environment Minister for a newly reunified Germany, I was absolutely convinced that such a development was impossible; but its passing has brought a lot of concrete, new problems. Unemployment, social problems—these are now the top priorities today, and therefore organising a conference, such as ECO 1997, is definitely more risky, because people now believe that the environment is number five or six on our priority list of political tasks.

I am totally convinced that this is a grave mistake. I am absolutely certain that the feeling and the knowledge and the conviction of the people in my country with regard to environmental policy has not changed. True, it is a little overshadowed by other current problems, but the long-term objectives are the same.

Let us not forget Ludwig Erhard (on 4 February in this year [1997] it was the 100th anniversary of his birth), who established the social market economy. My main aim was, is and will be to change the structure of the market economy to facilitate ecological progress. I am a market-oriented politician. It is my belief that we need financial instruments to stimulate development with regard to technology and social behaviour.

The most important problem in the past was that we subsidised the use of the environment's natural resources. The atmosphere was used without a price, as a dump for carbon dioxide and nitrogen oxides. Nobody investigated how to clean gases coming from a coal power station or from a car, because nobody was forced to invest money in it. So my most important aim is to integrate, wherever possible, environmental costs into the selling price of the product, with a particular focus on waste.

We have always had to struggle with the classic working division between one entity that is responsible for production, another for packaging, a third for selling, a fourth for consumption and finally, at the end of the chain, somebody responsible for waste. With such divisions, no one integrates waste into the concept of how to produce something, because that is not in their budget. The German Environment Ministry was constantly blamed for being only a recycling agency. My main objective was to integrate the handling cost of waste into the selling price of products, thereby creating an incentive to decrease packaging and to develop packaging materials that are easier, and therefore cheaper, to dispose of later on. If you are only a recycling agency, then forget it. You are only dealing with end-of-pipe technology and it is obvious now that that thinking has become completely obsolete.

Our challenge today is to start at the very beginning of the production of the product, not only with regard to packaging, but to the recyclability of the entire product as a whole. With this in mind, I was constantly in search of a solution for the disposal of end-of-life cars. We were not completely successful in this. I was aiming at a solution that required the car producers to take back their cars when they were no longer in use, when they became waste. The industry would then have had the incentive to build repairable cars that are subsequently recyclable.

Unfortunately, this life-cycle approach is very difficult to handle. We succeeded, when I was Environment Minister, in passing a life-cycle law, the *Kreislaus Wirtschaft Gesetz*. Of course, it was not easy to convince our industry to embrace such an approach. Nobody wants to take responsibility for the whole life-cycle because they are then accountable from cradle to grave.

The history of how we arrived at this point is revealing. In Germany, we had initiated a high-chimney policy. You didn't have to decrease emissions if you had a better chimney, a higher chimney, and you could do that very easily. So, for a long time in Germany, we simply tried to increase the height of the chimneys. And, of course, a similar approach was used from a horizontal point of view. I will never forget when I visited my colleague in Portugal. In those days he was very proud about the fact that the water was collected in a waste-water treatment station, which had a 3.5 km-long pipe going to the Atlantic. That was nothing more than a high chimney on a horizontal basis to better distribute and to decrease the specific pollution load.

The second phase was the end-of-pipe technology. We installed filters into the chimneys or we installed filters into waste-water treatment plant design. We were not eliminating the undesirable substance but concentrating it and rendering it easier to handle for landfill or other means of disposal.

The third phase was to look at how we could integrate all of this into the technologies themselves. We invented and put into place all of the labour-saving technologies imaginable, but no environment- or energy-saving policies, as these were not part of the equation back in those days. So we misplaced our resources in the intensive development of the wrong technologies.

When I was responsible for the CSD [Commission on Sustainable Development], I was always asked, 'How do we arrive at a better transfer of technologies?', and I always countered with, 'Do we have the right technologies in the developed countries for solving the problems in the developing world?'. I don't believe we do. To export our agricultural technologies, for example, to developing countries would be a disaster. In Germany, agriculture is the second most capital-intensive industry next to the chemical industry. So we have a very highly capital- and energy-intensive agricultural business. Is that really the right technology to export to a region where there are a lot of people in need of work? So I have to ask, 'Can we not only filter out but also integrate?'. Isn't it possible to create a new technological environmental framework? I think this is linked to the question of an ecological and social market economy.

As Minister of Urban Affairs, Construction and Planning, I am in a situation where I can try to act within the integration model. Let us look, for example, at cities. I was the head of the German delegation to Habitat 2 in Istanbul last year [1996] in June. It is an absolute must for an environmentalist to concentrate on cities, because cities will be the future all around the world. We are not heading, as Marshall McLuhan said we would, towards an urban village, but to an urban agglomeration.

More than 50% of the population will live in mega-cities. How to organise those cities with regard to sustainability, economic, ecological and social questions is the most difficult challenge at hand today. For example, the main problem of traffic in the city cannot be solved merely by better modes of transport but must be solved by improved infrastructure. How do we change this?

In the 1930s European architects agreed on a model that separated different functions of the city from one another. But, if you segregate, you obviously produce traffic problems. What we need now is to create the city of short distances, to integrate everything again. I have discussed it with architects, with city planners, and the ideal would be to create something like a new charter of ethics, a new model, a completely new paradigm.

If you achieve better integration, you decrease traffic needs. But we are now facing the Americanisation of our cities. To have retail shops located on the highway outside the city is of no benefit to the city centre. It only decreases the function of the city centre resulting in social and environmental problems. So we need to integrate environment into city planning.

And, then, what is the right regional structure? Of course, German history offers a better model for regional structures. We don't have, as is sometimes suggested in France, 'Paris and the desert'. We have a federation of states with their own capitals. Munich is the capital of Bavaria and Bavarians are proud of it. So we have a better demographic structure. We have to do our utmost to maintain this. Bringing the government to Berlin, for example, makes it imperative to move other institutions from the city. Our Environmental Protection Agency in Berlin has decided to relocate to a city in eastern Germany. If we don't continue to do this, very soon the difference between the east and the west in Germany will be like the difference between night and day, .

The difference between developing countries and developed countries is that in developing countries the slums are on the outskirts of the cities and in developed countries the slums are in the centre of the cities. If you go to downtown Washington, DC, or any other American city, you will see this phenomenon. I really want to alter this and create sustainable cities. In making Berlin the capital of the reunified Germany, I want to do my utmost to ensure we do not create a kind of government and parliament ghetto there. I want the government agencies to be integrated into the city. To accomplish this we decided to use only existing buildings. That is acting environmentally, because it is, in effect, recycling buildings, and there is a direct correlation—the older the building the more appropriate it is for recycling. It is wonderful, and easy, to make new use of fine old Prussian buildings. The prefabricated buildings of the 1950s, on the other hand, are a nightmare to change and redesignate.

So I want to integrate the city. I hope that we can be successful in this aim, especially with regard to housing. I have to construct 12,000 housing units for people relocating from Bonn to Berlin, but we want to avoid creating 'the Bonn centre in Berlin' where we would have a virtual fence around it with all the Bonners living there. So we have identified more than a hundred different locations in the city in which to integrate them into the community.

Then we face the problems of globalisation and homogenisation with regard to regional identity. If we lose this regional identity, I believe that it will be a destabilising factor in overall development in the world. Therefore—and don't misunderstand me, this is not nostalgia—we absolutely must do our utmost in Europe to maintain the genuine profile of the different regions and countries. If we misinterpret European unification as unification also with regard to culture, production and whatever, it would be misdevelopment on a very grand scale.

I believe that the question of how to integrate society will unfortunately be more difficult in the future. We are beginning to see a more telematic way of life where we are not necessarily linked with our neighbour if we don't want to be. I believe that this is one of the most serious problems we will face in the future. There is a high-level advisory board to the Secretary General at the United Nations, and, under the leadership of a former colleague, Birgit Adal from Sweden, they submitted a telling report about what may cause the disintegration of societies in the future. We now have not only global products and global production processes; we also have global information. Wherever you are, you have the same information as everyone else so you have a kind of unification of information around the world. You can have 100 channels on the television, but they are all very similar.

Conversely, I think there is something of a renaissance of regional feeling. The further you go in the globalisation of the world, the more people are looking for their home, their neighbourhood. In Germany there is renewed interest in regional dialects. Years ago this was virtually unheard of. I think this focus on regional identity is also vital for environmental policy. There is now interest in what is going on, for example, with my local river, or with the biodiversity of my region.

There is a story about a holiday camp, where young German and French students were given a test. The results showed that, with regard to theoretical questions about carbon dioxide and what damage it causes and so on, the Germans were much better, but with regard to practical questions, such as 'What kind of a tree is this?', 'What makes up the typical landscape?', the French were much better.

This story illustrates an important point: changes must be made not only at the top level but at all levels of society, so that everyone feels responsible for their environment, for their neighbourhood, for nature. And in doing so we will create new customers for the products that will be produced in the new sustainable industries of the future.

Such changes are, however, difficult for governments to impose. For example, we are now increasingly subsidising mobility, with the result that cities are directly linked to subsidised prices. I always emphasise that the structure of today is the result of the prices of yesterday and the prices of today determine the structure of tomorrow. Of course, we have discussed this very intensively in Germany—increasing petrol prices, for example, was one possibility. The opposition asked for an eco-tax to increase the price of energy, especially fuel, up by roughly 20 pfennig. That is, of course, a big increase for those people living in rural areas needing a private car. Therefore we need some kind of step-by-step approach. I learned a lot from our British colleagues. They decided to increase petrol prices year by year in 5% increments so that, after 5–6 years, the price of petrol will be significantly higher. If you know this, you can plan for your next car to be a smaller one, or for your petrol use to decrease, or to live elsewhere.

I am therefore proposing incremental solutions so that everyone has a clear signal of what is coming. All the wonderful ideas must be visible in a democratic society to convince people that this is the right way to go. To that end, we have decided, together with the Berlin Senate, that we will aim for a limit of 18%–20% of private transport in the centre of Berlin, with 80% commercial.

But my main thought today is that we are not signalling strongly enough to people that the environment will be the key issue in the next decade. We must stick to this conviction, and I am very glad that we are now co-operating with industry. I am totally convinced that those industries not looking to produce environmentally friendly products will lose their markets in the future and we must continue to send the right signals to industry to develop new products or production processes.

In conclusion, I repeat the essentials gleaned from my own personal experience. I started my business life in corporate economic policy. I have been in Brazil; I have been in Malawi, for example, in the southern part of Africa. The most important thing I learned was to ensure that people have their own identity, that they are genuinely integrated, and I believe that, if we cannot overcome the consequences of homogenised globalisation, it really would be a disaster. For business, the integration of regional issues into global strategy offers great opportunities to develop different types of products and contacts with your customers.

1999

Today, as we said in Rio and more recently in Malaysia, we are witnessing a new kind of cold war. This time it is between the industrialised countries of the North, where the forces of industrial development exist, and the countries of the South. In fact, costs linked to the environment have always been globalised, but the economic returns are in the North. The desertification of Mauritania is the example of a least-developed country hit first by industrial pollution (from mining, whose profits went elsewhere) and now with extreme weather conditions, caused by global climate change. Such distancing, or externalisation, of environmental problems brings to mind the vision of a global commons, a future where fundamental changes will have to occur—perhaps even changes in the notions of public/private property.

The fact that UNEP's world headquarters are in Nairobi reminds us daily that:

- 65% of the 3.5 million inhabitants of Nairobi live in shanty towns.
- In such a setting, it is impossible to isolate the environment as a factor in development and we see that the most toxic environmental waste is poverty.
- The gap between the rich part of the world and the least-developed countries is widening, and the opening-up and globalising of our economies most certainly contribute to this.

The solution lies in development that is both environmentally friendly *and* poor-friendly.

ZERO EMISSIONS

'Zero emissions' is a strong metaphor for change: an idea that has gained wide following in Japan. The zero emission approach makes use of life-cycle assessment and is an excellent example of 'circular, not linear' thinking.

The United Nations University in Tokyo has a zero emissions research initiative looking at the parameters and implications of this concept. The basic concept holds that, instead of considering waste as a problem within one single industry, it can become an opportunity or a resource for another production process. With the clustering of industries to take advantage of these opportunities, transportation costs and lag times in usage are eliminated. Two decades ago, ideas about cleaner production within individual companies contributed to the awareness that the manufacturing sector needs to improve its performance for its own interests. In moving on to the idea of circular or zero emission thinking, industry becomes important in the change to a sustainable society.

The following 1997 paper by Hiroyuki Fujimura (President, Ebara Company, Japan), describes their innovative zero emission concept which deals with industrial ecology, where wastes are redefined as inputs, providing a practical challenge for any engineer or industrial designer. Ebara moved into environmental engineering by way of pump production, then by water and air treatment.

ZERO EMISSIONS
An environmental engineering firm's challenge

Hiroyuki Fujimura

An approach to global environmental problems

In modern society, we have been pursuing economic development that is based on mass production, mass consumption and mass disposal. As a result, we are now facing serious environmental problems. In order to overcome these problems, industrialised countries must strive to reduce their environmental loads by changing their social and industrial structures and building compatible social and industrial systems.

Developing countries must review both good and bad examples set by the industrialised countries in order to avoid making the same mistakes. Environmental preservation and energy conservation must be placed as top priorities, and long-term development plans must be made, and followed, carefully. Co-operation between industrialised countries and developing countries is vital, and also beneficial to all. Technical assistance including some technology transfer from industrialised nations is a key element here. It will stimulate the change in society.

Responsibilities of environmental engineering firms

Japanese industries have begun their efforts to reduce environmental problems. These efforts will be most effective if they are co-ordinated. Governments may play the role of leader, providing the industries with vision and basic direction. For example, the Industrial Environment Vision, drawn up by the Japanese government, outlines the framework for environmentally friendly industrial activities, based on new developments in environmental issues.

Zero emission industrial systems

Zero emission industrial systems are composed of two levels of systems. They are:

1. Clusters of manufacturing processes with minimal environmental loads
2. Clusters of industries linked for net minimum waste and maximum energy conservation

Zero emission approach

On top of waste minimisation efforts through conscious choices in daily lives and industrial processes, including recycling, strategic clustering of industry groups are formed for further waste utilisation. Waste from one industry is used as raw material or feedstock for the other.

At the first level of the zero emission approach, individual industries must minimise their waste by:

1. Using 'clean energy' with low environmental load
2. Using 'green materials' which are raw materials that can be obtained with minimal disturbance of an ecosystem
3. Using 'clean materials' which produce minimal waste and environmental impact when processed
4. Improving industrial processes in order to achieve maximum process efficiency with resource and waste reduction
5. Recovery, re-use and recycling of energy and materials
6. Reassessing the fundamental product design with life-cycle assessment (LCA)

Of course, it is also important for industries to make profits. However, we believe that improving processes and cutting energy requirements and waste output will contribute to overall cost reductions.

Now, let us move from individual industries to clusters of industries. In the zero emissions approach, webs of industries are established for even further waste minimisation at the global scale. Waste to one industry may be gold to the other. Keeping the material in the cycle contributes to energy and other resources conservation, and to waste reduction as a whole. The optimum results would be obtained by employing LCA.

An environmental engineering firm's challenge

We can compare an industrial infrastructure to the human body. In Japan, we often refer to the business sector that deals with the treatment of industrial wastes as the 'veins of industry'; and the business sector that supplies required energy and materials to the rest of industries as the 'arteries of industry'. Ebara Corporation has many years of experience in end-of-pipe waste treatment. Based on this experience, Ebara is now also involved in developing the conversion technologies necessary to provide engineering services for supplying the 'arteries' with the valuable energy or raw materials that are recovered from the 'veins'.

The Fluidised Bed Gasification Process, presently being developed by Ebara, is an example of efficient conversion technologies. This system is vastly superior to the traditional systems in thermal energy recovery efficiency and reduced impact to the environment. Granted, this gasificator still cannot be called a 100% zero emission system because of the CO_2 release from the stack. These releases, however, are significantly less than releases associated with CO_2 off-gassing from the waste, had it been landfilled. Also, there are promising technologies being developed to reduce the CO_2 emission. An example is a red diode big-reactor. Other chemical absorption methods are being studied as well, producing promising results.

In addition to thermal recycling, a resource recovery system can be integrated into this gasificator. Besides metals (such as iron, copper, aluminium and nickel), valuable substances such as hahnium, alcohol, methane, gasoline precursors and ammonia may be recovered from municipal and/or industrial waste. Residual ash can be melted together and utilised as a construction material, or as an input for a cement. Furthermore, the gasification process effectively eliminates hydrochloric acid at a low temperature and the attached melting process completely destroys hazardous chemicals such as dioxins and polychlorinated biphenyls (PCBs) with a reduced air ratio at a high temperature.

An advanced pressurised gasification/melting furnace makes it possible to produce ammonia from municipal and/or industrial waste. Ammonia is of value to several industries.

With progressive development of efficient conversion technologies and systems, establishing various zero emission industrial systems is becoming a reality. Below are some examples of zero emission systems that may be realised in the near future.

Agriculture is one of the main industries in China. There is a great demand for cheap and productive fertiliser, and we expect that the demand will continue to rise in the future. In 1992, China produced approximately 20 million tons of fertiliser. This is a 230% increase from the quantity produced 15 years earlier. Now the government plans further increases of production, to 30 million tons a year by 2000. Presently, about 80% of electricity is generated through fossil fuels, especially cheap and abundant coal. This trend is expected to continue. However, coal-fired power plants produce oxidised sulphurs and nitrogen, a main cause of acid rain, which has a detrimental effect on the environment.

However, based on the zero emission concept, Ebara has come up with an innovative conversion system which we call the 'Electron Beam Fuel Gas Treatment System'. This system solves the gas emission problem while contributing to the production of fertiliser. By adding ammonia to the power plant furnace effluent gas, and then irradiating the electron beam, ammonium nitrate and ammonium sulphate are produced as by-products. Ammonia needed in this system can be obtained from municipal and/or industrial waste. The fertiliser produced from the flue gas and waste can now be used as a valuable material in agriculture. In other words, the system takes care of waste while helping to produce food for people. It becomes a kind of zero emission system, which aims at promoting economic development while conserving nature and energy.

Economic feasibility of zero emission systems

Introduction of zero emission systems into our society is beneficial for our present and future environment. Economic feasibility studies have shown that zero emission systems can be feasible under the right conditions.

For example, introduction of the electron beam (EB) system in China seems promising. Assume that we built a system in China for a power plant that produces 2 million Nm^3 per hour of flue gas.[5] If the price of ammonia is ¥13, and the price of the nitrogen-based fertiliser is ¥10/kg, the break-even point for this EB plant is roughly estimated to be 10–15 years. While the EB system was originally built for desulphurisation of the power plant flue gas, it can also produce profit as a fertiliser manufacturing system.

The ammonia to be used in the EB system may be obtained from car shredder dust that is treated by a conversion process with the gasification/melting furnace. Shredder dust, as we know, is yet another industrial waste. In order to produce 1,000 tons/day of ammonia, it would take about 900 tons/day of shredder dust and 600 tons/day of coal. If we get ¥20,000/ton as shredder dust disposal fees from its producers, this system will pay for itself in several years. Therefore, while this system takes care of the shredder dust, it also functions as an industrial ammonia-manufacturing process.

These examples show that there could be a profitable and greener future with zero emission strategies.

5 Nm^3 is a normal cubic metre, with 'normal' meaning a temperature of 25°C and a pressure of 1 atmosphere.

ZERO EMISSIONS IN CONSTRUCTION

The Taiheiyo Cement Corporation, headquartered in Japan, is the second largest cement producer in the world (5,120 million tons a year). Whereas cement and concrete facilitated the construction of the 20th century's infrastructure, today the industry's, or at least Taiheiyo's, mission is the realisation of symbiosis with nature. Taiheiyo is thus also engaged in advancing the concept of 'zero emissions'.

Michio Kimura, now President of Taiheiyo, joined the Nihon Cement Corporation in 1955 after taking his law degree at Tokyo University. He moved into management in 1982 and then on up through the company, which merged in 1998 with Chichibu Onoda Cement Company to form the Taiheiyo Cement Corporation. He delivered the following presentation in Paris in 1999.

ZERO EMISSIONS

Clustering of industries (industrial ecology in practice)

Michio Kimura

'Industrial clustering' is herein defined as the formation of a cluster of different industries in which waste or by-product from one industry is used as a raw material or fuel by another industry within the cluster. It is one of the best countermeasures for reducing the environmental impact caused by industrial activities.

Japan consumes 1,950 million tons, or 12%, of the world's natural resources while having only 2% of the world's population. Of the 1,950 million tons, 35%, or 700 million tons, is imported from overseas. Overall, Japan generates a total of 450 million tons of waste of which 400 million is industrial waste from the processing of natural resources into industrial products and 50 million is urban waste from the consumption of those resources and products. The quantity of waste does not seem to be decreasing despite reduction efforts. So far, however, 39% of the waste is re-used or recycled and 61% is pre-treated by de-watering, drying and incineration, thus reducing the total waste by 41%, leaving 20% or 90 million tons per year for landfilling.

Being a relatively small mountainous country, Japan faces a serious shortage of landfill sites. This, together with an increase in illegal dumping, is causing serious social and environmental problems. The remaining available capacity of existing landfill sites is estimated to be sufficient for only another six or seven years, which is extremely worrying considering that the NIMBY[6] syndrome makes it virtually impossible to build new sites. For example, the estimated cost to build one new landfill site in the Tokyo Metropolitan Area, with all the safety specifications required for public acceptance, is around US$500 million.

Under such dire circumstances, the Japanese cement industry plays a crucial role. With its 25 plants located evenly over Japan, the industry easily consumes

6 See footnote 3 on page 24.

27 million tons or 6% of the nation's waste in the form of fuel and cement raw materials. As a result, on a global scale, the Japanese cement industry boasts the smallest energy consumption per ton of cement. In addition, by using recycled materials, the industry reduces carbon dioxide emissions by 16% compared to when using only virgin materials. Therefore, a successfully functioning cluster has already been formed around the cement industry in Japan.

It is notable that the Japanese electrical power industry's 53 thermal power plants, each with a 500–1,000 megawatt capacity, cannot function without the cement industry. Desulphurising calcium carbonate for the cleaning of the power plants' exhaust gases is supplied by the cement industry and, perhaps due to the 'karma of clustering', the resulting spent calcium carbonate residue is gypsum, which is a necessary raw material for cement production. Also, 50% of the fly ash produced by the power stations is used as a cement raw material. Bulk transportation mechanisms such as tanker vessels and shuttle trains carrying fly ash and calcium carbonate have been exclusively set up to continually commute between power plants and cement plants. Therefore, industrial clustering among power plants, cement plants, and also plasterboard plants, is successfully in progress.

As previously stated, cement industries in Japan utilise 27 million tons of waste a year, of which 11 million tons is utilised solely by Taiheiyo Cement Corporation. More than a hundred types of industry are clustered around this one company, including power generation, chemical, steel, non-ferrous, paper and automobile industries, as well as oil refineries and many municipalities and prefectures.

Through experimenting with new methods of waste re-use and with the help of international knowledge, we have made some fantastic leaps forward. For example, the problems resulting from widespread use of chlorofluorocarbons (CFCs) in terms of ozone layer depletion are well known. However, no one knew how to destroy the CFCs. Soon after the 'Cement Rotary Kiln Method' was accepted in the Montreal Protocol, by entering into contracts with Tokyo Metropolitan government, Saitama Prefectural government and Mercedes-Benz Japan, Taiheiyo Cement Corporation was the first in the world to perform successful experiments and to establish a viable method for the destruction of CFCs.

The cement industry of Japan is competitively developing technology to receive more waste from other industries, municipalities and prefectures. It has embarked on a mission to become the biggest contributor to zero emissions. The ultimate goal is to produce cement without using virgin raw materials and without using virgin fuel. This is not a dream but a realistic possibility.

Most municipal waste and sewage sludge is incinerated in Japan due to the lack of available landfill sites. Since incineration ash, and especially fly ash, contains harmful dioxins, heavy metals and chlorine, it cannot be recycled as a raw material for conventional cement. But with the newly developed eco-cement technology, a creation of Taiheiyo Cement Corporation, waste plastics, including PVC (polyvinyl chloride), and other industrial wastes that are difficult to use in the conventional sense as raw material or fuel, can now be utilised. Dioxin in the ash is decomposed by the high temperature inside the cement kiln. Heavy metals are extracted by wet-type refining technology to later be recycled for the non-ferrous metal industry. The Public Research Institute of the Japanese Ministry of Construction has estimated that, for eco-cement production, CO_2 emissions will be cut by a staggering 50% and energy consumption will be reduced to 89% compared to that of conventional cement production. The decision to construct eco-cement plants in both the Tokyo Metropolis and Chiba Prefecture has been finalised. The Tokyo plant will produce 160,000 tons of cement per year with the waste from 4.3 million people, while the Chiba plant will produce 100,000 tons per year with the waste from 2.5 million people.

The amount of waste generated throughout the world during the last 200 years is estimated to be around 100 billion tons. This waste is being increased further by the explosive growth in both population and economy in developing countries. Since waste destroys natural ecology, we believe in establishing an industrial ecology, or second ecology if you will, to separate waste from natural ecology. This will be accomplished by the recycling of waste from one industry as a raw material or fuel for another. In other words: industrial clustering. The cement industry can be seen as the hub of the cluster.

The Japanese government provides much aid to developing countries through environmental ODA (overseas development assistance). We propose that, in order to be more effective in the long run, governments should encourage and implement industrial clustering while providing ODA. For example, the gift of a construction package to a developing country could consist of a power plant, an oil refinery, a paper plant, a cement plant, a gypsum plant, a sewage disposal plant and a municipal incineration plant. Alongside existing manufacturing plants, this could form a whole industrial cluster with all the features entailed therein, such as energy savings, resource savings, waste recycling and so on—the first realisation of a zero emissions society in the developing world.

When faced with environmental matters, the problem with Japanese industries is their focus on flow production, excessive quality and technical optimism. A flow production orientation inevitably results in wasteful resource consumption and large-scale waste generation. Also, a demand for excessive quality results in a lower efficiency of raw materials and production energy. Although a new perspective for our materialistically oriented civilisation is necessary, we are optimistic that it is technology that will solve our environmental problems in the future. Nuclear weapons, land mines, chemical weapons, PCBs (polychlorinated biphenyls), DDT (dichlorodiphenyltrichloroethane) and CFCs were all developed by technology.

INDUSTRIAL SYMBIOSIS

'Industrial symbiosis' is similar to zero emissions and also to industrial ecology, and is, perhaps, more successful as a metaphor than the latter, simply because 'industrial' and 'ecology' would appear to be oxymoronic, whereas it is easy to imagine a symbiosis among diverse industries.

The successful industrial symbiosis operating in Kalundborg, Denmark, has become a world-famous model for industrial ecology developments elsewhere—the classic system where the wastes of one enterprise become the raw materials of another. Interestingly, the symbiosis at Kalundborg evolved accidentally and grew organically, of its own accord.

Erling Pedersen, who is Danish, is Chief Executive Officer of the Industrial Development Council in the Kalundborg Region. Pedersen started out as a Danish government customs official who moved on to administration for a civil engineering firm in Greenland, then to financial controlling for the Danish Army Material Command before finding his niche in municipality administration. He now possesses a unique experience in developing networks and co-operation between municipalities and private companies, and Kalundborg is a unique partnership for change. Pedersen spoke of the extraordinary Kalundborg experiment in Paris, in 1999.

REMARKS (1999)

Erling Pedersen

The industrial symbiosis in Kalundborg, Denmark, located 100 km west of Copenhagen, is an environmental network that has developed over the past two decades. It was not a planned network, but a series of projects initially quite independent from one another. There was no original joint management, but rather bilateral agreements between independent partners. And, most interesting, the network did not evolve with any academic knowledge of scientific environmental network theories, but as good and economical management practice. The original incentive to all the 'shared' projects at Kalundborg was profitability. All projects required investments and resulted in revenues or savings for the parties involved.

What is the industrial symbiosis? It is an environmental co-operation between five industries and the municipality of Kalundborg, an example of industrial ecology that recycles water and waste, and transfers energy. The partners of the industrial symbiosis are:

- Gyproc, a plasterboard factory
- A/S Bioteknisk Jordrens, a company undertaking remediation of polluted soil by using microorganisms
- The 1,500 MW Asnaes Power Station, the biggest in Denmark
- The Statoil Refinery, also Denmark's largest with an annual production of 4.8 million tons of oil products
- Novo Nordisk, whose plant in Kalundborg produces insulin and industrial enzymes
- The Kalundborg Municipal Department for water and energy

The first project was completed in 1982, and involved the transfer of a supply of steam from the power station to Novo Nordisk and to Statoil via foundations built and used also by the municipality. This agreement taught four partners how to co-operate in a new way.

Then, in the 1990s, the Statoil Refinery had to build a waste-water plant. The quality of the treated water was so good that the power station could use it for cleaning purposes. In 1993, the power plant had to take measures against the emission of sulphur. The smoke had to be cleaned. A method was chosen by which the smoke is treated with chalk lime (calcium hydroxide). Gypsum is formed and can be utilised as a raw material for the production of plasterboards at Gyproc. This gypsum has replaced natural gypsum imported from Spain where it is mined in environmentally unfriendly open mines. The 'industrial' gypsum has a uniform quality and performs very well.

Eventually, almost accidentally, an environmental consciousness developed even though profitability is still the principal incentive. Now, our objectives are also the lowering of resource consumption and the reduction of emissions to the air, water and soil. At first, projects simply emerged. Now, we actively try to discover new possibilities. Clearly, communication and confidence are important factors for the realisation of the economical and environmental potential that lies in an inter-company network of co-operation.

We believe that the total investments in all of the 19 projects amounts to approximately $75 million. The present annual savings for all partners is $15 million, and the total savings/revenues accumulated so far are $160 million.

Environmental benefits and resource savings

Total water consumption

The companies participating in the industrial symbiosis—the industrial symbiotes—have together reduced their water consumption by about 1.4 m^3 or 25%. Asnaes Power Station also now re-uses its own waste-water.

Oil

Together the participants have reduced their yearly oil consumption by 19,000 tons. An 'eternal' flare of surplus gas is part of the safety system at all refineries. At Statoil, the flare is reduced to a night-light because both Gyproc and Asnaes Power Station use the surplus refinery gas as fuel instead of using oil or coal. The gas from Statoil replaces 90% of the oil consumption for plasterboard production at Gyproc.

Coal

Coal consumption has been reduced by 30,000 tons a year. Approximately 2% of Asnaes Power Station's coal consumption has been replaced by surplus refinery gas from Statoil.

Carbon dioxide (CO_2)

By replacing some coal with refinery gas, Asnaes Power Station has reduced its CO_2 emissions by 30,000 tons a year, or 3%.

Sulphur dioxide (SO_2)

Emissions of sulphur dioxide have been reduced by 25,000 tons a year, or about 58%, through flue gas desulphurisation at Asnaes Power Station and Statoil. The by-products of this process are gypsum at the power station and sulphur at the refinery.

Fly ash

Asnaes Power Station has an electro-filtration unit for removing ash from the flue gas. Every year, the building and cement industry uses about 170,000 tons of this fly ash for production of cement and concrete.

Gypsum

Gyproc receives 80,000 tons of gypsum a year from the Asnaes Power Station, corresponding to about two-thirds of its requirement. The gypsum replaces natural gypsum in the company's plasterboard production.

Sulphur

A local company, Kemira, uses 3,000 tons of sulphur from Statoil each year to make sulphur dioxide.

Biomass

Enzyme production at Novo Nordisk is based on a fermentation process which produces nitrogenous biomass; 1,500 tons of nitrogen and 600 tons of phosphorus in the form of biomass are piped or transported to local farmers to fertilise their fields.

INDUSTRIAL ECOLOGY

Brad Allenby, Vice President of AT&T, has written and spoken widely on industrial ecology, which is about putting the 'eco' into industry as well as into eco-nomics. A father of the industrial ecology field, Dr Allenby is again at AT&T after a stint as Director, Energy and Environmental Systems at the Lawrence Livermore National Laboratory, USA, where he was in charge of a $250 million research programme. The interest of large government and industrial organisations in industrial ecology is clearly high. Is industrial ecology thus becoming a new science? Regardless of what it is called, according to Allenby it is apparent that substantial fundamental research is required if sustainability and sustainable development are to become more than slogans. Few people, however, even recognise the profound depths of our ignorance, much less understand the need for an integrated, international research and development programme. The following is a 1997 paper that Allenby presented in Paris.

IS INDUSTRIAL ECOLOGY A NEW SCIENCE?

Brad Allenby

It is important to realise the fundamental nature of the shift of environmental issues from overhead to strategic for consumers, producers and society, which is the essence of sustainability. As professionals, as industry, as society, we have yet to recognise the challenge posed by this shift, or how profound our ignorance really is. It is not an insult to current research efforts in such areas as design for environment (DfE) or life-cycle assessment (LCA) to say that we know almost nothing yet about sustainability. The research necessary to support such change has really not yet begun in earnest. Two examples might be useful to illustrate this point.

1. There is much discussion about 'sustainable communities' and 'sustainable companies'. If, as is likely, sustainability is an emergent characteristic of a properly self-organised complex system—that is, the global economic structure—these terms are oxymoronic: sustainability is a property of the whole, and not of the parts. The terms are useful in that they indicate a generic goodwill towards the environment. To a scientist, however, they beg the question. One must begin by asking: what is the physical and energetic basis of the community or firm in question? What stocks and flows of materials support the community? What is the linkage between these processes and supporting natural systems? What are the environmental impacts embedded in products and materials imported into the communities, which have the effect of exporting the community's impact around the world? How do different communities compare along these dimensions? Although some sporadic, high-level work has been done on these issues, the real task has yet to be begun.

2. If a manufacturer were to ask a designer to design a 'green telephone', she would of course say yes, as few are actively against the environment.

> Her first question, however, would be: 'what are the preferred materials for the various applications?' Such data do not now exist. If one cannot answer even such a simple, reasonable question, how far are we, then, from understanding sustainability?

In considering these issues, it is important to first construct an intellectual framework, which supports a reasoned, rational approach. As materials choice, use and management over the life-cycle are obviously important aspects of improving the environmental efficiency of economic activity, their relationship to levels of the intellectual framework are used as illustrations. It is important to note that our ability to understand what 'sustainable material use' might be is virtually non-existent, but we can certainly make progress at the DfE/LCA level in improving the environmental efficiency of current practices and technologies. Fundamental research is not, in other words, a substitute for incremental progress—or a reason to avoid making today the improvements that are possible. Research at all levels of the system can, and should, proceed simultaneously.

In brief, it is obviously difficult to sketch out an entire research agenda. Many obstacles to progress in such research exist, including the fact that reductionist scientific approaches, which is the standard in Western countries, are of little value when the perturbations are profoundly systemic. Some progress in defining the needs is, however, being made. In fact, there is a group of industrial ecology experts in the United States, known as the Vishnus, that have been working on just such an agenda under the sponsorship of Lawrence Livermore National Laboratory. This must be regarded as preliminary work, however, and of course it does not begin the research actually required.

Two points bear emphasising. First, our ignorance in this area is profound and vast; most people, even researchers, are unaware of the extent to which our technologies, our science, our cultures must change if we are to approach sustainability. Second, whether it is called industrial ecology or something else, a massive scientific and technological research and development effort will be required to support such an evolution. Such an understanding will not be sufficient in itself to achieve a sustainable state, but it is an absolute prerequisite to it.

INDUSTRIAL ECOLOGY IN FRANCE

Odile LeCann is Executive Director of France's Comité 21, a non-governmental organisation whose membership is made up of representatives from large French businesses, agencies, municipalities, research groups, associations and, increasingly, medium-sized businesses. Comité 21's agenda is to create a dialogue and provide information about the integration of environmental concerns into all aspects of French life. Their agenda is creating true sustainable development in the 21st century.

Comité 21 and the French community of Grande-Synthe, in collaboration with the utility giant, Gaz de France, have spearheaded the creation of an industrial ecology zone in France. The feasibility study for the project is by Suren Erkman, internationally known author on industrial ecology. The goal is to target candidate industries for the zone on the basis of their complementarity.

LeCann, who was director of Comité 21 when she wrote this paper in 1999, was engaged in developing local Agenda 21 programmes for French municipalities, and in optimising environmental considerations that are resulting from international deregulation of the energy industry. At this crucial turning point, governments are encouraged to invest in alternative energies or energy-delivery systems, rather than in nuclear or in the fossil fuels that powered the industrial world born in the 19th century.

INDUSTRIAL ECOLOGY IN PRACTICE
The French case

Odile LeCann

France has the good fortune to be in a favourable position regarding CO_2 emissions, largely because of the dominance of nuclear power as an energy source. The short distance France must go to meet the Kyoto Protocol standards of 2008/2012 is all the more daunting. Yet France is far behind in implementing industrial ecology or zero emission policies.

Nevertheless, French industrialists have practised various kinds of industrial ecology strategies for at least 20 years. For example, since the 1980s, Nestlé France has marketed used coffee grounds (a fermentable waste) as compost to mushroom growers in Saumur. The same group also marketed the caffeine removed from coffee in the production of decaffeinated coffee to the manufacturers of Coca-Cola. Makers of cognac in south-western France currently use the biogas emanating from production residue as an energy source for electricity and heating, in a manner similar to 50% of the country's water purification plants.

Thus, industrial biocenosis exists in France, but randomly, not as a policy of economic development or of town planning.

There have been various reasons for this non-start in industrial ecology zero emissions as public policy. First, unlike Germany, the Netherlands or Japan, France still has sufficient open space for widespread industrial development. This will certainly not continue indefinitely. Second, it was only in 1982 that political decentralisation created any kind of real power, including judicial, at the local level. Local communities suddenly found themselves dealing with new industries as part of both their revenues and demand for services; industries were faced with a new level of regulation. Yet there is still little dialogue between the networks representing industry and the networks representing local communities.

Third, zero emissions as a policy is seen by industrialists as a constraint since the move, at the European Commission level, from BATNEEC (best available technology not entailing excessive costs) to BAT (best available technology). This is especially true of industries that have not integrated environmental considerations into their practices. Some examples are: (a) the 1996 Clean Air Act, which imposes emission standards on atmospheric pollution, urban transport planning and regional air-quality planning; (b) new regulations regarding engine and turbine emissions, which will cost certain industries billions of francs within five years; (c) the request to all prefectures from the Minister of Planning and Environment for new Departmental Waste Elimination Plans integrating recycling and take-back. Further, the European directives regarding persistent organic pollutants (POP) emissions, auto oil and others make French industry sceptical and reactive.

Fourth, industrialists seem uninformed when it comes to the international conventions related to zero emissions. When you speak to French industrialists regarding the international climate convention, or about biodiversity or desertification, they envision the United Nations or distant governmental groups, far from the realities of the field. For example, the Buenos Aires talks of 1998 were sufficiently complex and technical that industrialists felt sidelined and were therefore negative. Unfortunately, this lack of information can lead to a lack of preparation. In February, 1999, the World Bank held a conference in Paris on Carbon Prototype Funds. There were few industrialists present and no small business representatives at all.

In spite of everything, new national, European or international opportunities are likely to eventually stimulate the French industrial sector toward zero emission practices. What might these be?

First, the arrival of new procedures, such as the ISO 14000 series or local approaches to Agenda 21. In 1997, only 11 French businesses were ISO 14000-certified. Today, there are more than 200. One of the first was Lexmark, the manufacturer of laser printers, who discovered that certification played a role in obtaining export contracts, particularly to northern Europe. Many industrial groups are finding that ISO 14000 brings with it a competitive gain. In October 1997, the Intergas Marketing Association stated that sustainable development approaches are essential for competitiveness among users of natural gas and that this is now a question of competitiveness, not ideology. For small towns and communities, the development of Agenda 21 guidelines can proactively stimulate environmentally friendly new industrial development.

Second, the gradual understanding of the mechanisms of international pollution rights. These mechanisms are at the heart of the international climate control convention. Tradable emission permits will regulate industrial development, the growth of certain countries, and technology transfers between developed and non-developed nations. Already people speak of 'tons of carbon emissions per inhabitant'. There might even be a sort of stock exchange dealing with such pollution rights. These mechanisms could provide French industry with new markets in technology transfers, if industrialists seize the available opportunities provided by the World Bank, the International Monetary Fund (IMF), and other funds. There is much at stake, and little has been done so far.

Third, the greatest stimulant in France for the development of industrial ecology zones is the opening of energy markets. In France, there have been only two groups creating, importing, transporting and distributing energy: EDF (Electricité de France) and Gaz de France. Two European directives were approved in 1999 opening the French market for certain large industrial energy users. In the first stage (by 2005), 33% of energy consumption by large industrial groups can be opened to Europe-wide competition. In the second stage, local towns and communities will have the same opportunity. As a result, energy providers are becoming multi-energy and multi-service, including diversification into telephone services, water services and telemanagement of buildings. There would appear to be an opening here for eco-industries and industrial ecology zones in new kinds of partnerships.

Fourth, the determination of towns and communities to commit to local Agenda 21 guidelines is an advantage in the development of zero emissions zones. As certain older types of local economic development (ZI,[7] ZAC,[8] etc.) haven't been well received, there is an opportunity to create new objectives in development zones.

A feasibility study for an industrial ecology zone is now under way in a city in the north of France. First the community developed an Agenda 21, then a sustainable development goal within the various municipal services. This kind of positive approach to developing such a zone can ultimately give a welcoming example to concerned industrial sectors. Here is an example: the Clean Air Act of 1996 requires the installation of atmospheric pollution monitoring equipment—first in all municipalities larger than 100,000 inhabitants, and then, by 2001, in all towns. Here, then, are ready markets and opportunities for collaboration between industries seeking to develop technologies for communities not yet prepared for such a move.

New markets connected to zero emissions can be profitable, if considered from a point of view combining technical, economic, ecological and social aspects. But the leaders of all these diverse networks must meet and work together.

In the Promethée programme in Bourgogne, established through the Regional Council, some industries saved 30% of their costs simply by auditing the flow of their primary materials and by finding markets for their waste. Almost more importantly, these same industries began working with regional institutions and other area industries, many of which turned out to be fellow subcontractors to even larger groups. One can easily imagine the new structures of collaboration that are possible. France may well be passing out of the zero emissions observation phase and into the proactive phase.

7 'Zone industrielle', industrial zone.
8 'Zone d'aménagement concerte', zone developed in conjunction with local stakeholders.

MONITORING WHAT MATTERS

Another potential metaphor for change is 'monitoring what matters' which includes 'green' gross national products (GNP) or 'genuine savings' at the level of the economy, or environmental performance indicators and 'green accounting' at the level of the individual business. The GNP is now known to be an inadequate measure for monitoring overall progress. The development of 'indicators' for monitoring what matters is continuing in the environmental field. For example, an attempt to produce a sustainability indicator was presented by the Global Leaders for Tomorrow Environment Task Force to the World Economic Forum in Davos, early in 2000. This sustainability indicator attempts to integrate 64 variables into one 'Pilot Environmental Sustainability Index'.[9]

US Senator Robert Kennedy had the following philosophical take on monitoring what matters 30 years ago:

> Too much and too long, we seem to have surrendered community excellence and community values in the mere accumulation of material things. Our gross national product . . . if we should judge America by that—counts air pollution and cigarette advertising, and ambulances to clear our highways of carnage. It counts special locks for our doors and the jails for those who break them. It counts the destruction of our redwoods and the loss of our natural wonder in chaotic sprawl. It counts napalm and the cost of a nuclear warhead, and armored cars for police who fight riots in our streets . . . Yet the gross national product does not allow for the health of our children, the quality of their education, or the joy of their play. It does not include the beauty of our poetry or the strength of our marriages; the intelligence of our public debate or the integrity of our public officials. It measures neither our wit nor our courage; neither our wisdom nor our learning; neither our compassion nor our devotion to our country; it measures everything, in short, except that which makes life worthwhile.[10]

There are many statistical, practical and political problems in 'monitoring what matters' at the level of the economy and society, but some countries are now developing 'headline indicators' that try to capture the essence of changes in 'wellbeing' or in 'quality of life'. Recent proposals for headline indicators from Germany, the Netherlands and the UK are summarised in Table 2, taken from the European Environment Agency's 1999 *Environment in the European Union at the Turn of the Century*.

9 For further information, contact www.weforum.org, or the Yale Center for Environmental Law and Policy, New Haven, CT, USA, at epcenter@pantheon.yale.edu.

10 'Recapturing America's Moral Vision', address, University of Kansas, Lawrence, KS, 18 March 1968.

5EAP themes	***Environmental policy themes and indicators (Netherlands)***	***Environment Barometer (Germany)***	***Sustainability counts (UK)***	**Gröna Nyckeltal *(Sweden)***
Climate change	Climate change (greenhouse effect and depletion of ozone layer) Index based on emissions of CO_2, CH_4, N_2O and production of CFCs and halons	Climate: CO_2 emissions	Climate change: emissions of greenhouse gases	Climate change: emissions of CO_2
Acidification and air quality	Acidification: indicator based on depositions of SO_2, NO_x and NH_3	Air: emissions of SO_2, NO_x and NH_3	Air pollution: days of air pollution (urban and rural sites)	Acidification: emissions of acidifying substances (NO_x and SO_2)
Urban environment	Disturbance: percentage of Dutch people affected by noise and odour in Neq			Urban air quality: benzene levels in the atmosphere (winter half-year mean value in various urban areas)
Waste management	Waste disposal: index based on the total quantity of solid waste dumped annually			Waste: waste for landfill (deposited quantities of waste material in Sweden)
Management of water resources		Water: percentage of flowing waters at which the mandated goal of chemical quality class II for AOX and total nitrogen is achieved	Water quality: rivers of good and fair quality (percentage of total river length)	
Coastal zones	Eutrophication: index based on emissions of phosphates and nitrogen to soil and water			Coastal areas and eutrophication: load of nitrogen and phosphorus into the sea
Protection of nature and biodiversity		Nature: ecological priority areas (absolute and as percentage of non-settled area)	Wildlife: populations of wild birds (index)	Nature: protected forests (as a portion of productive forest land)

5EAP = 5th Environmental Action Programme; CO_2 = carbon dioxide; N_2O = nitrous oxide; CFCs = chlorofluorocarbons; NO_x = nitrogen oxides; NH_3 = ammonia; SO_2 - sulphur dioxide; Neq = noise equivalents; AOX = adsorbable organic halogens

Table 2 **Synopsis of national sets of environmental headline indicators** *(continued over)*

Source: EEA 1999a

5EAP themes	*Environmental policy themes and indicators (Netherlands)*	*Environment Barometer (Germany)*	*Sustainability counts (UK)*	**Gröna Nyckeltal** *(Sweden)*
	Toxic and hazardous pollutants: index based on the dispersion of agricultural pesticides, other pesticides, priority substances (cadmium, polyaromatic hydrocarbons, mercury, dioxin, epoxyethane, fluorides, copper), and radioactive substances			
Soil		Soil: increase per day in area covered by human settlements and traffic routes	Land use: new homes built on previously developed land (percentage)	
Resources	Resource dissipation	Resources—materials: resource productivity (GDP per ton of raw materials) Resources—energy: energy productivity (GDP per primary energy consumption)		Energy: use of energy (energy consumption related to GDP), electricity for heating
Transport			Transport: road traffic (vehicle-miles)	Transport: environmentally adapted means of transport (the portion of journeys to and from work and school taken on food, by bicycle or public transport) (private transport by car in kilometres per person aged between 6 and 84)
Industry/ agriculture				(Sustainable enterprises): number of environmentally registered enterprises (EMAS or ISO 14001) (Agriculture): recovery of phosphorus in sludge to agriculture

5EAP = 5th Environmental Action Programme; EMAS = EU Eco-management and Audit Scheme

Table 2 (continued)

Meanwhile, at the level of an individual business, there are many innovative ways to monitor progress with environmental and social impacts, as well as with profits: that is, the 'triple bottom line' (see Part 3, 'Tools for change': 'Financial tools', page 219, and 'Environmental reporting', page 234). The World Business Council for Sustainable Development (WBCSD) has developed criteria and indicators for monitoring progress with eco-efficiency at the corporate level.

A highly technical text on the subject of indicators (which are how monitoring what matters happens) follows, crunched by experts from the European Environment Agency.

DEVELOPMENTS IN INDICATORS
Total material requirement (TMR)

European Environment Agency

Indicator	*Policy issue*	*DPSIR*	*Assessment*
Total material requirement (TMR)	Decrease burden to global environment due to resource extraction	Pressure	–
Domestic TMR	Decrease burden to domestic environment associated with resource extraction	Pressure	✓
Foreign TMR	Decrease burden to environment in foreign countries	Pressure	✗
Direct material input (DMI) versus GDP	Improve materials productivity (processed materials only)	Response	✓

✓ = positive; – = neutral; ✗ = negative; DPSIR = Driving forces–Pressures–State–Impact–Responses, i.e. the socioeconomic and environmental framework; GDP = gross domestic product

Table 3 **Developments in indicators: total material requirement (TMR)**

Extraction of natural resources in EU member states declined by 12% between 1985 and 1995, but imported resources increased by 8% between 1995 and 1997. The direct material input (DMI) in the economy fell by 8% on a per capita basis at the beginning of the 1990s, but then increased slightly. In most member states, economic growth has been associated with increased DMI. However, Finland, France, Italy and the UK have reduced their dependence on direct material input.

In past years, a number of aggregated physical measures have been proposed to show overall pressures on the environment. Examples include the 'human appropriation of net primary production' (indicating the share of biomass used by human activities in energy units) and the 'ecological footprint' (indicating the area of productive land utilised by a certain population and its activities). The new indicator, total material requirement (TMR), expresses the total mass of primary materials extracted from nature to support human activities. Thus, TMR is a highly aggregated indicator for the material basis of an economy.

The TMR indicator includes both materials used for further processing (DMI) and hidden flows: that is, extractions that are not used further, but have an environmental impact (e.g. overburden and extraction waste). TMR includes extraction both from domestic territory and of the resource requirements associated with imports. Changes in the balance between the foreign and the domestic amount of TMR indicate possible shifts in environmental impact between countries.

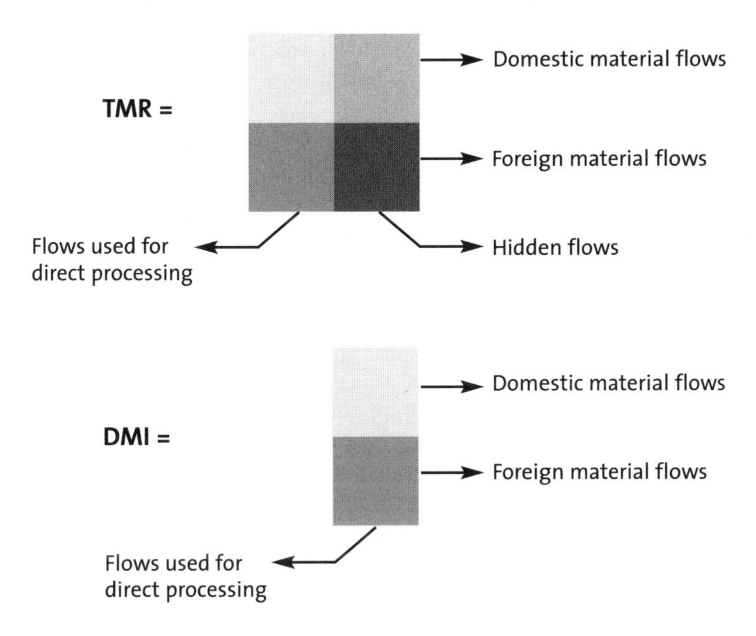

TMR = total material requirement; DMI = direct material input

Figure 19 **Total material requirement and direct material input**

TMR indicates a generic pressure on the environment. The volume of resource requirements determines the *scale* of local disturbances by extraction (e.g. devastation of mining sites, disruption of natural habitats, ground-water contamination and landscape changes at the extraction site), the throughput of the economy (DMI) and subsequent amounts of emissions and wastes. TMR, however, does not indicate the *severity* of these specific environmental pressures at the individual sites.

TMR consists of all resource extraction besides water and air. Statistics on industrial production, agriculture, forestry and fisheries provide data on domestic material requirements, while foreign trade statistics give data on imports (grouped into raw materials, intermediate products and final products). Raw materials are traced back to the delivering countries using these statistics. This data is supplemented by specific information on hidden flows, such as overburden and extraction waste in mining and quarrying, excavation during construction and dredging, and erosion of agricultural fields. Intermediate products are classified according to their main constituent (e.g. steel or aluminium) and combined with data on cumulative resource requirements. Final products are accounted for only by their weight. The resulting values therefore represent minimum estimates for the total material requirement.

TMR comprises all the primary resources needed for the production side of an economy, including trade and service activities. All inputs contributing to value added are considered: that is, pure transit is not accounted for. Countries with a high dependence on either domestic resource extraction or imports exhibit high TMR values, irrespective of whether the resulting produce is exported or consumed within the own country.

So far, TMR has been calculated only for a few European countries (Bringezu 1997; Adriaanse *et al.* 1997, 1998; Juutinen and Mäenpää 1999; Mündl *et al.* 1999). The first calculation of TMR for the European Union is presented and analysed below. Although the values are preliminary, the order of magnitude appears sufficiently valid for international comparisons.

Material resource requirements

In 1995, the EU's TMR amounted to 18.1 billion tons or 49 tons per capita (Fig. 20). Due to the masses of materials involved and their hidden flows, the EU's TMR is dominated by energy, metals and mineral resources. It is significantly lower than that for the US in 1994 (84 tons per capita), but higher than that for Japan (45 tons per capita) in 1994. Both the US and Japan have a higher GDP per capita than the EU. By comparison, GDP per capita for Poland was a fifth of that for the EU in 1995, but the TMR per capita was almost 60% of that for the EU.

Two non-overlapping time series are presented—one for the domestic and one for the foreign components in TMR (Figs. 22 and 23). In 1995, the domestic part was 63% of total EU material requirements, having fallen over the previous ten years. The remaining 37% of TMR was linked to imports; this value increased slightly between 1995 and 1997.

The main reason for the much lower total resource requirements in the EU compared with the US is the difference in material flows related to fossil fuels. Due to less use of energy in the EU and a reduced use of coal, Europe's fossil-fuel resource requirements are only 44% of the US's (Fig. 21).

Figure 21 also reveals differences in national patterns of material requirements:

- Because Germany still depends to a large extent on coal extraction, material flows related to fossil fuels are the same order of magnitude as in the US.

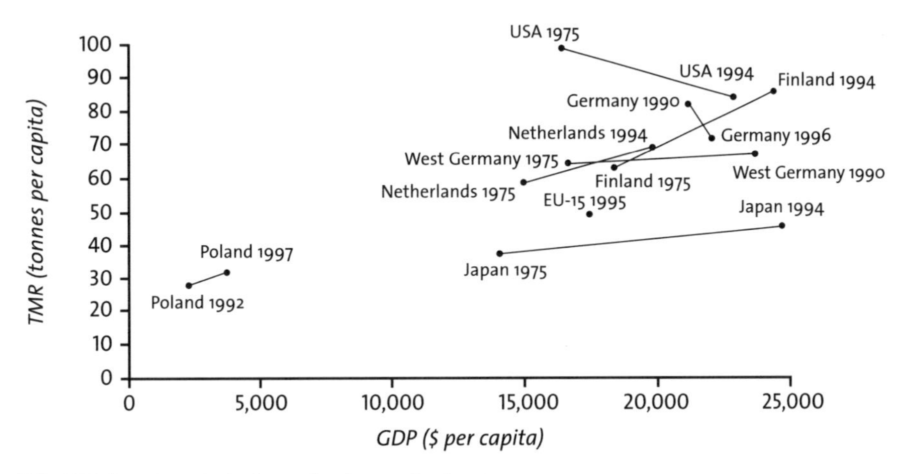

Note: GDP given at constant prices and exchange rates for 1990.

Figure 20 **Total material requirement (TMR) and gross domestic product (GDP) of the European Union compared with selected member states and other countries**

Source: Wuppertal Institute, World Resources Institute, National Institute for Environmental Studies, Netherlands Ministry for Housing, Spatial Planning and Environment, Thule Institute and Warsaw University

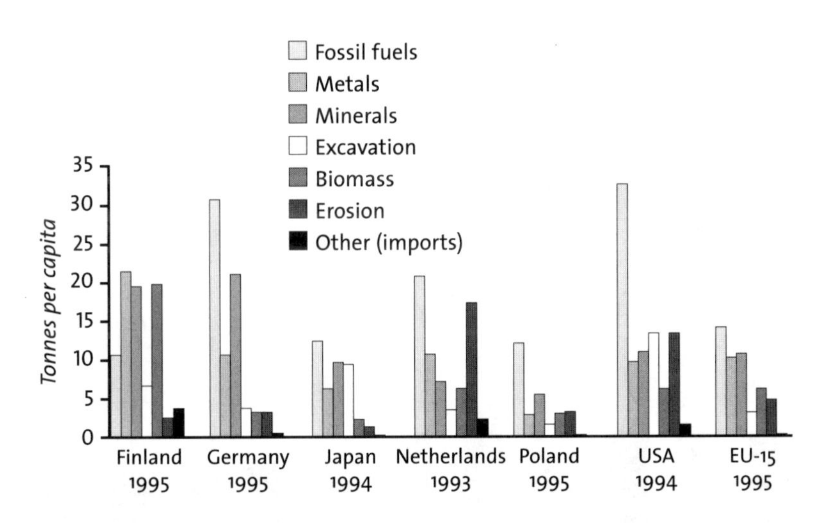

Note: hidden flows are included in fossil fuels, metals and minerals or are represented by excavation and erosion.

Figure 21 **Composition of total material requirement (TMR) in the European Union, selected member states and other countries**

Source: Wuppertal Institute, World Resources Institute, National Institute for Environmental Studies, Netherlands Ministry for Housing, Spatial Planning and Environment, Thule Institute and Warsaw University

- Germany and Finland have the highest rate of minerals extraction due to sand and gravel production. The German value for minerals is twice the EU's owing to significant housing and infrastructure construction.
- In Finland, where metal manufacturing is still a significant part of industrial production, resource requirements for metals are relatively high. The relatively high biomass values for Finland are due to forestry (timber is a significant Finnish export).
- The high material flows associated with erosion in the Netherlands reflect its significant agricultural imports from non-European countries.

Domestic resource extraction

The domestic portion of TMR for the EU fell by 12% between 1985 and 1995 to 63%, mainly due to a decrease in the extraction of fossil-fuel resources (Fig. 22).

The reduction was mainly due to a decline in lignite production following the closure of significant numbers of obsolete industrial facilities in eastern Germany since reunification. However, lignite production still represented 80% of fossil fuel-related domestic resource extraction in 1995 and was associated with 23% of

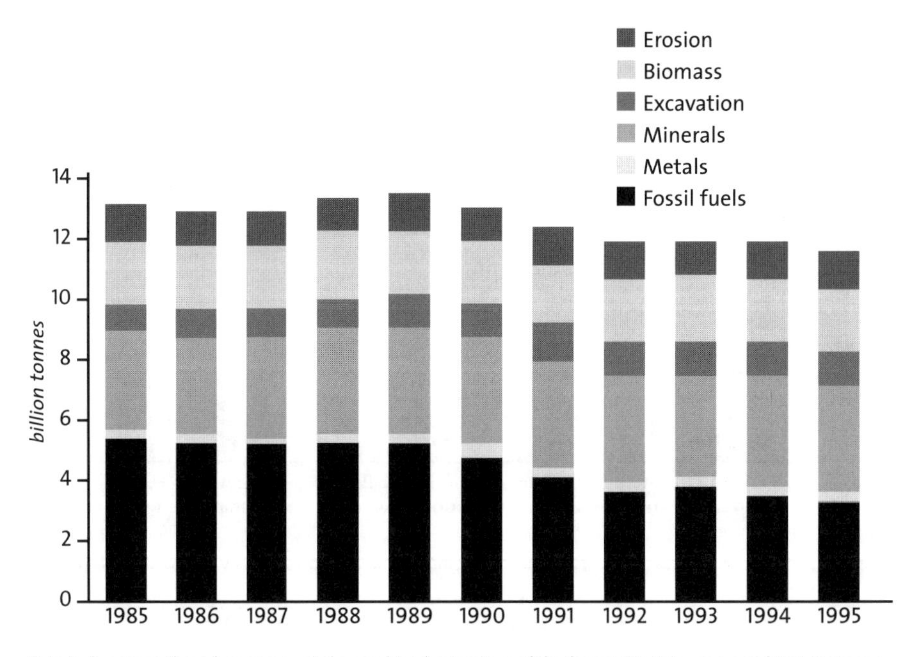

Note: before 1990, the values represent the combined extraction of the former West Germany and East Germany.

Figure 22 **Domestic resource extraction in the EU between 1985 and 1995**

Source: Wuppertal Institute

the domestic TMR of the EU. The main producers were Germany (74% of EU lignite production), Greece (21%) and Spain (4%).

Extraction of hard-coal resources declined less rapidly but still significantly compared with lignite—by 35% since 1985 to 135 million tons in 1995. In 1995, the main producers of solid hard coal were Germany, the UK and Spain with 44%, 38% and 13% respectively. However, in terms of total extraction (including hidden flows), these countries accounted for 35%, 24% and 39% respectively. Hard-coal production in Spain thus has much higher hidden flows than Germany and the UK.

The decline in energy-resource extraction was greatest for those energy carriers with the highest hidden material flows. For lignite, an average of nine tons of overburden has to be removed to extract one ton of the energy carrier. This ratio, which highlights the poor resource efficiency of lignite production, has grown gradually. For hard coal, the ratio is much lower (around 1:1), but is also increasing slowly. The ratios for the other energy carriers are significantly lower. As lignite and coal production decline, these highly resource-intensive energy resources are being replaced by the less resource-intensive oil and gas.

Simultaneously with this decline in domestic material requirements for fossil fuels, the volume of mineral requirements has grown and, in recent years, has exceeded domestic energy resource extraction (Fig. 22). Quarrying activities should therefore be taken as seriously as mining activities. The pressures on the environment associated with the overall extraction volume: for example, hydrological changes, habitat disturbances, growth of built-up area and construction waste have probably increased as well.

EU resource requirements in foreign countries

Imported metals, minerals and agricultural products are associated with higher hidden flows per commodity than domestic produce, indicating a relatively higher environmental impact in the exporting countries. In 1995, the resource extraction associated with EU imports was at least 37% of TMR. Between 1995 and 1997, it increased by 8% mainly due to the import of precious-metal ores (Fig. 23). Renewable resources account for only 2.4% of foreign TMR, compared with 18.3% of domestic TMR. Foreign TMR thus contributes particularly to the depletion of non-renewable resources.

- EU material requirements for foreign resources increased by 8% between 1995 and 1997. Demand for luxury and precious commodities has a major influence on foreign TMR.

Imports of precious-metal ores into the EU increased by 51% between 1995 and 1997 to 5,600 tons/year. In 1997, resource flows of precious metals (an estimated 1.5 billion tons) contributed around 70% to metal-resource extraction for imports to the EU while iron and copper ore, the second- and third-ranked metal imports,

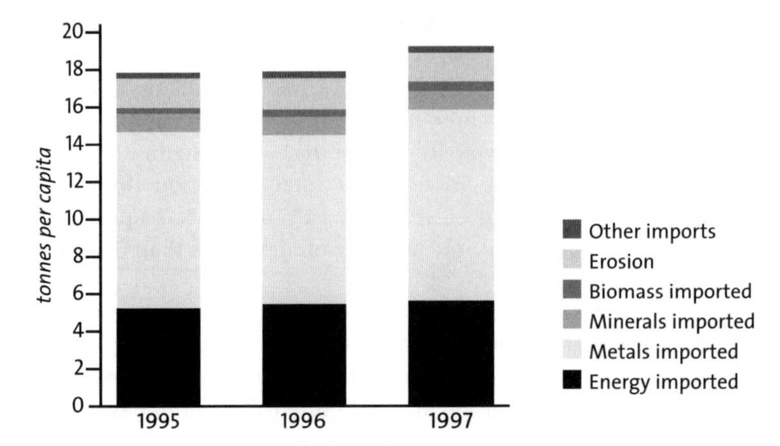

Note: foreign resource extraction as a basis for domestic activities.

Figure 23 **Total material requirement (TMR) of the European Union associated with imports**

Source: Wuppertal Institute

contributed only 18% and 4% respectively. Imports of finished products such as jewellery, plated ware, gold and silver goods also contribute to resource requirements. These have not yet been included in foreign TMR data for the EU, but are estimated to contribute an additional 1 ton per capita.

Diamond imports dominate mineral requirements. Imports of only 44,000 kg in 1997 were linked to an estimated extraction of 232 million tons of material. This is more than half the mineral-resource requirements of the EU's foreign TMR. The hidden flows associated with the import of 2,450 tons of other precious stones in 1997 have not yet been quantified due to lack of data.

The inevitable conclusion from the data above is that much of the resource flows for EU imports are associated with luxury commodities.

There is a marked difference in the hidden flow–commodity ratios for domestic resource extraction and foreign resource extraction (Table 4).

	Domestic	***Foreign***	***Total***
Fossil fuels	3.48	1.63	2.55
Metals	1.07	15.49	10.34
Minerals	0.21	4.41	0.31
Agricultural biomass	0.63	5.90	0.89
Total	**0.94**	**4.18**	**1.51**

Table 4 **Ratios of hidden flows to commodities for the European Union in 1995**

Source: Wuppertal Institute

Fossil-fuel imports (other than electricity) have a significantly lower hidden-flow ratio than domestic extraction of energy resources. Imports are mainly oil and natural gas, and have lower hidden flows than lignite and hard coal. Reducing energy use by industry, transport and households will mean less burden on the environment from resource extraction either domestically or in foreign countries.

Hidden flows from the import of metal resources are 14 times higher than those from domestic extraction. Ore mining is only a minor activity within the EU, which imports most of its base metals (iron, aluminium, copper, etc.) and almost all its precious metals.

Imports of agricultural products by EU member states are associated with more erosion than domestic agriculture. This is mainly due to the import of products such as coffee and cocoa. In a number of member states, consumers have shown some interest in supporting more sustainable agricultural practices by buying specific and labelled products.

Resource productivity of direct material inputs

Calculation of TMR requires connecting production and import statistics with coefficients for the hidden flows. Production of a time series of direct material inputs (DMIs), i.e. inputs of primary materials without the hidden flows, would be much easier and would give a straightforward and up-to-date indicator for trends in resource productivity. When comparing countries for which TMR and DMI have been calculated, there is an indication that a high DMI goes with a high TMR and vice versa. If such a correlation could be proved, the more easily calculated DMI could be used for regular monitoring of materials productivity. A full domestic TMR would then only need to be calculated if the burden of resource extraction to the national environment was required. In addition, foreign TMR can be used to indicate the sharing of burdens and the shifting of problems between countries and regions.

The DMI of the EU showed a moderate reduction in absolute terms of 6% between 1988 and 1995 (Fig. 24). On a per capita basis, it declined by 8% from 21.2 tons per capita to 19.5 tons per capita. Most of the change occurred at the beginning of the 1990s and was mainly due to a decline in imports of 1 ton per capita. Since 1993, however, the DMI of most EU member states has been increasing slightly. Thus, in terms of DMI, there is no sign of an absolute decrease in material use.

When comparing DMI and GDP for EU member states between 1988 and 1995, three groups of member states can be distinguished:

1. A higher economic performance is associated with higher DMI in Austria, the Benelux countries, Denmark, Greece, the Netherlands, Spain, Sweden and Portugal.

2. Germany and Ireland achieved significantly higher GDP with a constant DMI. In these two member states, a relative decoupling of direct material requirements and economic growth has occurred.

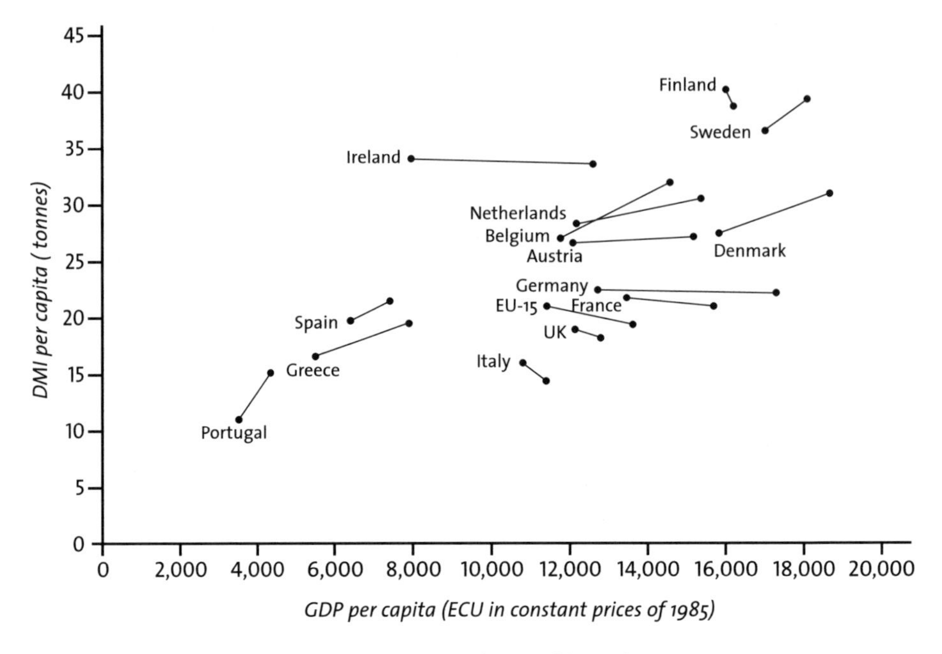

Note: DMI of member states includes intra-EU trade, but the DMI of the EU does not.

Figure 24 **Direct material input (DMI) versus gross domestic product (GDP) per capita in European Union member states, 1988–95**

Source: Wuppertal Institute

3. Finland, France, Italy and the UK managed to combine economic growth with reduced DMI. Reduced extraction of building minerals allowed these four member states to demonstrate that absolute dematerialisation is possible.

The EU as a whole performed well with a reduction of DMI per capita by 8% while GDP per capita increased by 19%. Altogether direct materials productivity grew by 29% in the EU between 1988 and 1995. The difference in EU results compared with those of individual countries is due to the exchange of goods between the countries: member states' DMIs include intra-EU trade whereas the EU's DMI does not. Due to the constant level of DMI since 1992, the EU as a whole can be grouped with Germany and Ireland (group 2)—leaving it the challenge of following the countries in group 1 to use less material resources while achieving a higher economic performance.

- Direct resource productivity of the EU increased by 29% between 1998 and 1995.

ENVIRONMENTAL DIPLOMACY

Environmental diplomacy is a relatively new development in foreign affairs and is not widely known outside the rather small group of its practitioners. The term reflects the recent integration of environmental concerns into globalisation and foreign policy.

Following are ideas from an article by Bettina Laville ('Le globe, l'équerre et le diplomate', *Annales des Mines*, April 1996), in which she defines environmental diplomacy. Laville was, at the time she wrote this, Environmental Advisor to the French Prime Minister Lionel Jospin.

THE RISE OF THE 'BIO-DIPLOMAT'

Bettina Laville

The United Nations Conference on environment and development held in Rio de Janeiro in 1992 gave to international diplomats new and even subtly revolutionary tasks that were perhaps not recognised as such by their governments.

Certainly, as we often hear, nothing will be the same after Rio. But it seems that the debate these days focuses more on the globalisation of the world's economic projects than on the internationalisation of policies assuring its survival. Nevertheless, little note has been taken of the fact that negotiations on environmental questions have left a particular mark on diplomacy. Diplomats in charge of the Rio follow-up appear in some ways to be guardians of things invisible: fragile biodiversity or climate stability. Like the scientist and visionary Ptolemy describing the shape of the Earth without really knowing it, today's diplomats are asked to fix limits to the modern world's negative effects on the Earth, and this with sometimes weak or even contradictory mandates from their governments. They are often asked to arbitrate between science and politics while displeasing both, even though politicians are often secretly relieved to leave to diplomacy the burden and the precautionary long-term responsibility of securing conditions for future wellbeing.

As a result, international diplomats are always looking for the most reliable information available concerning current physical phenomena, and the way scientists are used in this process places them under terrible pressure. First, there is the pressure exercised by the industrial sector which, because of economics, resists and shapes to its own logic any technological change demanded by greater planetary survival.

And, even though scientists are now present in worldwide summits such as those on climate change, they are often sidelined when it comes to the time for definitive negotiations or decision-making, because of the make-up of national delegations or the limited financial or diplomatic means of member countries.

But the scientists who play only a consulting role during actual negotiations thus lose their stakeholder role, even though one of the accomplishments of Rio was to make them stakeholders. It is therefore diplomats who must become diplo-scientists and who must consider complex constraints such as scientific data, economic impacts and geopolitical accords. Instead of a techno-science, it is a diplo-science that Rio has given us. Diplo-scientists must be 'bio-diplomats' who master, in the case of climate change, scientific knowledge and industrial interests, and all of this under the vigilant eye of large multinationals.

Is this to say that environmental diplomacy is a multidisciplinary diplomacy, but nonetheless one confined to expert debates which remain inaccessible to the public and sometimes even to government officials? Just as in the 16th century, when scientists understood the world better than the politicians who were governing it, today the actors affecting global change reflect better the state of our world than practitioners of classic diplomacy.

ENVIRONMENTAL DIPLOMACY IN THE US

Further thoughts on the subject and practice of environmental diplomacy were provided in a 1997 talk in Paris by Ambassador Mark G. Hambley, United States Special Representative to the Commission on Sustainable Development and Special Negotiator on Climate Change. Mr Hambley describes the US State Department's inclusion of the environment into its foreign policy. His point of view is both official and idealised, and does not address the fact that the United States government finds itself increasingly in a position of diminishing importance in terms of environmental leadership, due to the unwillingness of the US government to take any positive action in regard to global warming. Further decisions made during the current US administration have further undermined US leadership and often left its diplomats with no clear positions.

ENVIRONMENT AND SECURITY

Mark G. Hambley

Until relatively recently, environmental concerns took a decidedly second place in the hierarchy of issues attracting the attention of our country's diplomats. Not any more. Nowadays, the importance of the environment has come to be recognised as a key priority for governments, both domestically and internationally. That the world needs to find a better way to sustain its finite resources was at the crux of the United Nations Conference on the Environment and Development in Rio de Janeiro in June 1992.

Since Rio, there has been a significant increase in the level of both multilateral and bilateral diplomatic efforts on environmental issues. Three conventions flowed forth from Rio—on biodiversity, desertification and climate change. Five follow-on world summits were launched—Cairo on population issues, Copenhagen on social issues, Beijing on women's rights, Istanbul on issues related to habitat, and Rome on food security.

In addition, the Commission on Sustainable Development (CSD) was created by the UN General Assembly as a direct result of the Rio conference. This organisation has engendered scores of meetings in cities throughout the world. Many of these meetings have been held at the technical experts level, but diplomats are heavily involved in the annual plenaries of the CSD which attempt to promote progress on such issues as sustainable agriculture, technological transfer to developing countries, ways to finance sustainable development, and sustainable forest management.

Environmental issues are now in the mainstream of American foreign policy. There are now global environmental issues which our diplomacy must address in order to preserve a world that is both healthy and sustainable for future generations.

Environment and diplomacy: four areas of concentration

Forests

Forests are a key environmental issue which have acquired considerable diplomatic attention. As the world's largest producer and market for wood products, and the leader in biotechnology, the United States—like many countries—has an enormous stake in the sustainable use of the world's forests. Forest depletion has serious repercussions for global warming (they are huge carbon sinks), biodiversity conservation (they harbour untapped genetic resources), and agriculture (they prevent erosion and siltation).

American diplomacy has been a strong proponent of the UN Commission on Sustainable Development's consideration of a wide range of forest issues under its Intergovernmental Panel on Forests. In concert with other key government agencies, US diplomats have also taken several direct actions in high-priority areas to help promote sustainable use of forests on a bilateral basis. From Eastern Siberia to Suriname, we have participated in efforts to assist local governments to plan the rational use of their forests, often in the face of pressure to trade short-term financial incentives for ecologically damaging, clear-cutting deals in many of the world's remaining virgin forest areas.

Marine pollution

Marine pollution affects directly the health of fishing stocks worldwide. Pollution of the marine environment is caused by the deliberate dumping in the ocean of wastes, activities on land such as agricultural and industrial run-offs, sewage discharge and vessel discharge.

These issues are addressed in a number of global and regional fora. Vessel discharge issues are addressed in the International Maritime Organisation; dumping is regulated under the Global London Dumping Convention; and, following the 1995 Washington Conference, there is now a global programme of action for addressing the land-based activities.

Chemicals

The use of certain toxic chemicals and pesticides (such as DDT [dichlorodiphenyltrichloroethane] and PCBs [polychlorinated biphenyls]) in areas throughout the world is an increasing health threat, both to the people who use them and to individuals far from the area in which they are used. Because this poses a long-term health and environmental threat to all countries, we have placed a high priority on developing international agreements to regulate the trade, production and use of the most hazardous of these chemicals and pesticides, also known as persistent organic pollutants (POPs). We are pleased that the UN Environment Programme has agreed to sponsor negotiations beginning in early 1998 on a global convention to deal with POPs (EEA/UNEP 1998).

Climate change

Perhaps the leading environmental issue confronting the world today is the question of global warming or 'climate change', as the problem is more accurately described. Given the international nature of the climate issue, and the need to involve all regions of the world in the solution, the Department of State has the lead in the inter-agency process and for the implementation of policy in the multiple negotiating sessions taking place each year.

The Department of State is committed to taking the lead in each area related to our international environmental concerns. This will be done by bolstering our ability to blend diplomacy with science and to negotiate global agreements that protect our health and wellbeing.

As part of this process, we are incorporating environmental planning into each of our bureaus and designating key embassies as environmental hubs to address region-wide natural resource and environmental issues. These regional hubs will help co-ordinate US overseas efforts, and work with national governments, regional organisations, NGOs (non-governmental organisations) and the business community to identify environmental priorities. To effectively cover the globe, the Department plans to establish a total of 12 hubs. The first two hubs, in Amman, Jordan and San José, Costa Rica, opened in the autumn of 1996; four more by the summer of 1997.

In short, the environment has very much become a key watchword for modern-day American diplomacy. We have long defined threats to the nation's economic wellbeing as security concerns. Retaining access to certain markets, protecting the sea lanes, and ensuring access to economically important resources have long been key security priorities. Certainly climate change, ozone depletion and biodiversity loss—with their attendant impacts on US agriculture and other significant economic sectors—should be security priorities as well.

Part 2

Partnerships for change

Change in environmental issues forces many stakeholders to re-evaluate their relations with others—with governments, scientists, customers, suppliers, financers, employees, local communities and the public. Often, stakeholders are trying to establish new objectives and patterns of working together that are more in tune with the requirements of eco-efficiency, 'ecological space' and 'ecological footprints', 'right- to-know', environmental justice, environmental, social and ethical reporting, consumer pressure and Internet campaigning.

The texts included in 'Partnerships for change', while describing impressive progress in industrial practices, do not represent significant change in society's power relationships. Those new developments that do have the potential to radically change power relations are in Part 4, 'Civic actions for change'. The boundaries between these two kinds of change are not always clear, but there is, in any case, a difference between rearranging the deckchairs on a sinking ship and saving the ship.

PARTNERSHIPS WITHIN INDUSTRY

In Japan in 1996, the Keidanren (Japan Federation of Economic Organisations) announced voluntary action by Japanese industry directed at conservation of the global environment in the 21st century. The Keidanren was established in 1946 and now represents more than 1,000 Japanese industrial groups. The organisation plays an essential role in developing a consensus within the industrial community on important questions, and the following 1997 paper illustrates the close working relationships that are the rule within the structure of Japanese society. Then Chairman of the Committee on Environment and Safety of the Keidanren, Yoshifumi Tsuji (also at the time Chairman of the Nissan Motor Co. Ltd) came to Paris to present the Keidanren's statement regarding voluntary action. Yoshifumi Tsuji is now a Vice Chairman of the Keidanren and a Counsellor at Nissan.

THE KEIDANREN APPEAL ON ENVIRONMENT

Yoshifumi Tsuji

Five years have passed since the Keidanren instituted the Global Environment Charter. During this period, we have deepened our concern about environmental protection and taken positive steps, both at home and abroad, to tackle the matter. Nonetheless, the environmental issue, including global warming, has become increasingly serious in recent years.

For instance, countries concerned are required, under the Framework Convention on Climate Change, to target the stabilisation of the gross carbon dioxide volume at the 1990 level in the year 2000, but the volume has instead shown an uptrend in Japan. As for waste disposal, the Law for the Promoting of Sorted Collection and Recycling of Containers and Packaging has been enacted as a step towards the establishment of a recycling-based society. The realisation of such a community, however, calls for a basic change in the mode of conception, a change that targets 'resources' or 'by-products' instead of 'waste'. On the other hand, there is a growing international mood for environmental management and auditing, with the ISO 14000 series, a voluntary international standard in the private sector, scheduled to take effect in the autumn of 1997.

With the 21st century dawning, it is the hope of everyone to hand a well-protected environment and its blessings on to the next generations. We should restructure the 'throw-away civilisation' that leads to the waste of resources and achieve 'sustainable development' so as to meet the needs of the present without compromising the ability of future generations to meet their own needs.

As key words to employ in our efforts in that direction, we therefore attach importance to these three goals: (1) reconfirmation of 'environmental ethics' for individuals and organisations to honour; (2) realisation of 'eco-efficiency', a factor needed to reduce the environmental load through improved technology and economic efficiency; and (3) tightening of 'voluntary efforts' to cope with the environmental issue.

Motivated by this concept, we declare that, in the spirit of the Keidanren Global Environment Charter, which states that tackling environmental problems is essential to corporate existence and activities, we will take a voluntary, resolute and responsible approach in dealing with important tasks existing in the environmental field.

In facing up to these problems, we essentially need partnership with companies, consumers, citizens, non-governmental organisations (NGOs) and the government. Everyone should be well aware of being a 'global citizen'. So should every company be aware of being a 'global corporate citizen' and act in concert with the government, consumers, citizens and NGOs.

In order to awaken the people to such necessity, it is effective for enterprises to promote education on the environmental problem and positively tackle environmental enlightenment activities both inside and outside companies.

In the hope of thinking and acting together with the government, consumers, citizens and NGOs as 'global corporate citizens', we will transmit this declaration through the Internet and ask for others' views on our programme. We intend to reflect the comments and opinions thus expressed in mapping out industry-wide voluntary action plans aimed at protecting the global environment.

Measures for four urgent issues

Measures to cope with global warming

Making it a basic policy to review the 'throw-away economy', structure a recycling-based society and improve energy efficiency and carbon utilisation efficiency, we aim to maintain the world's paramount level of environmental technology. We also aim to improve energy utilisation efficiency on a global scale through transfer of appropriate technology to developing countries.

Concrete methods

1. Preparing industry-wide voluntary action plans incorporating definite goals and steps towards enhancement of energy efficiency, and periodically reviewing the progress of such actions
2. Recovery and utilisation of heat exhausted from cities and industries, reduction of natural energy costs, improvement of utilisation efficiency of fossil fuels through co-generation and compound generation, and the safe, effective utilisation of atomic energy
3. Improvement of energy efficiency through inter-industry collaboration based on the life-cycle assessment (LCA) concept
4. Co-operation in coping with global warming in the residential and commercial sector through development of energy-saving products
5. Positive participation in 'activities implemented jointly' to transfer technology to developing countries in close co-operation with the government

6. Promotion of forest protection and reforestation projects in developing countries through business corporations themselves and the Keidanren Nature Conservation Fund

Structuring of a recycling-based society

In order to review the throw-away-type economic community, where resources are liable to be wasted, and convert it into a recycling-based society, we will work on 'cleaner production' designed to attain optimum efficiency in all the processes from product design to disposal. At the same time, we will revise the conventional concept of 'garbage' and treat waste as a valuable resource, transcending the boundaries of individual industries. We will thus address recycling as the most important task in corporate management and make a systematic approach towards reduction of waste and recycling.

Concrete methods

1. Controlling the incidence of waste and re-utilising it from the viewpoint of life-cycle assessment (LCA) and developing products with full consideration given to the degree of recyclability and disposability (e.g. review of the frequency of product restyling)
2. Disposal of waste products by appropriate methods
3. Structuring systems for recovery and disposal of waste products
4. Use of waste products as raw materials by developing waste disposal technology through inter-industry collaboration
5. Simplification of packaging and promotion of recycling
6. Positive introduction of products with lesser environmental load and recyclable products

Restructuring of environmental management systems and environmental auditing

We will structure an environmental management system in an effort to address the environmental problem voluntarily, ensure its continuous improvement and perform internal auditing to confirm that the system will steadily work. Keidanren has positively participated in the formulation of the ISO environmental management and auditing standards scheduled to come into effect in 1997. It is recommended that Japanese industries, manufacturing or non-manufacturing, should utilise the standards as an effective means of environmental improvement.

Concrete methods

1. Prompt introduction of environmental management and auditing systems into corporations (e.g. appointment of officers in charge of envi-

ronmental problems, creation of an environmental department and enforcement of internal auditing)

2. Implementation of environmental management and auditing in conformity with the ISO standards or taking steps that correspond thereto

3. Playing an active role in the making of environmental labelling, assessment of environmental performance and LCA international standards under ISO

Environmental considerations in evolving overseas projects

International business activities by Japanese enterprises, such as overseas production and developmental imports, are rapidly spreading from the manufacturing industry to banking, physical distribution and service sectors. We will pay closer attention to the environment in stepping up and diversifying business activities overseas, as well as observing the 'Ten-Point-Environmental Guidelines for the Japanese Enterprises Operating Abroad' incorporated in the Keidanren Global Environment Charter.

In conclusion, we reaffirm the importance and urgency of every industrialist to be a 'global citizen', and express also as citizens our determination to make innovations in our lifestyle towards the goal of 'sustainable development'.

PARTNERSHIPS BETWEEN GOVERNMENT AND BUSINESS IN JAPAN

In Japan, the integration of governmental and business policy is a clear example of partnerships between governments and business. Following are excerpts from a 1997 talk given in Paris by Katsuo Seiki (then Executive Director of the Global Industrial and Social Progress Research Institute, MITI [Ministry of International Trade and Industry], Japan) about finding the right balance between command and control and voluntary agreements. The MITI is the most powerful ministry in Japan. It is the MITI that finances, develops and co-ordinates, nationwide, Japanese industrial policy, a policy that has traditionally drawn its strength from its long-term approach which emphasises innovation. MITI developed a document called *The Environmental Vision of Industries*, which shapes Japanese industry, sector by sector, according to life-cycle thinking. Instead of a fragmentary approach, this integrated approach seems to put Japan at the forefront of sustainable business strategic development, which requires close partnerships between government and business.

JAPAN'S ENVIRONMENTAL POLICIES

Katsuo Seiki

Japan's new environmental policies

What we need today is for both government and the private sector to address new environmental problems with their unique features and to modify policy instruments and measures accordingly.

First, environmental issues are increasingly getting across national borders, and becoming 'globalised'. The typical cases include the climate change issue related to the emission of greenhouse gases such as CO_2, the ozone depletion issue caused by atmospheric release of CFCs (chlorofluorocarbons), and the issue of cross-border hazardous waste. Unlike traditional pollution issues, where the adverse effects are normally limited to factories discharging pollutants and the surrounding area, newly emerging issues affect wider regions across borders or globally. It is essential, therefore, to intensify international co-ordination and co-operation efforts to resolve and mitigate such issues. It means that, in addition to efforts already exerted by developed countries, additional international collaboration will be required between developed and developing countries and between developing countries.

The second characteristic is the longer time-frame. Although some degree of lead time did exist in case of past pollution problems between pollutants' discharge and their hazardous consequences, current environmental issues, or those issues expected to intensify in severity in the future, would emerge or materialise over a much longer time-frame. In the case of the CO_2 issue, for example, its century-long impact makes it difficult to fathom how the CO_2 emitted today would affect human society in the future. It is quite conceivable that, by the time any sign of hazardous effect arises, it will be too late to introduce any effective mitigation measures. In this effect, it will be extremely important to use the precautionary principle for implementation of policies and measures.

The third feature is the involvement of a wider range of human activities as a cause of environmental problems. While traditional pollution followed a simpler formula of a manufacturing plant being the source of hazardous materials, with surrounding community residents as victims, current emerging environmental issues paint a more complicated picture involving a wider range of residential and commercial activities. Take the CO_2 issue, for example: any single entity may be emitting CO_2 and, at the same time, it is a victim of its global impact. Therefore, it is necessary to introduce and promote environmental measures to cover every entity of economic activity, including citizens, businesses, national governments and local administrations, all sharing roles equally and fairly.

Fourth, many of the aforementioned environmental issues begin to introduce the concept of the life-cycle of products and services. Concerning the major role of business in the economy, it will be necessary for business to incorporate environmental concerns at each phase of a product cycle, such as raw material procurement, manufacturing, distribution, sales, consumption and waste disposal, whereas the conventional pollution concept requires the reduction of the environmental burden in the manufacturing process only.

To address such environmental issues and to stipulate the principles of environmental policies, the Japanese government established the Basic Environmental Law in 1993. Its basic philosophy was to build a sustainable society with less environmental burden. For this, it urges every state and local government, every corporation, every citizen to integrate environmental concerns into daily activities, with a special emphasis on voluntary measures. It also requests national governments to implement necessary measures, such as the institution of a technological assistance programme, etc.

The Environmental Vision of Industries

MITI (Ministry of International Trade and Industry) has developed environmental measures based on the concept put forward in the Basic Environmental Law. It already introduced policy instruments such as regulatory measures, taxes and subsidies, as well as the measures to promote voluntary actions by industries to address environmental issues. Since no single policy instrument or few of them can provide a key to resolve any issues, MITI created and implemented a comprehensive package of policy instruments.

To promote voluntary actions by industries, MITI introduced various relevant programmes such as those seen in the establishment of guidance for voluntary actions, the institution of eco-labelling, and a campaign to enhance public awareness.

The Environmental Vision of Industries, introduced in June of 1994 by MITI, focuses on the industrial sector, a major participant in economic activities. Its purpose is to promote voluntary actions by industries by setting forth guidelines to illustrate what environment-friendly actions can be integrated in their business activities. The guidelines have been published and circulated to encourage public awareness of environmental issues. The details of the Vision are as follows.

First, environmental impacts were analysed and identified for every phase of a product's life-cycle in 15 major industries. Life-cycle phases were raw material procurement, manufacturing, distribution, sales, consumption, waste disposal and so on. The 15 industries include the manufacturing industries of iron and steel; aluminium; non-ferrous metals such as copper, lead and zinc; processed metals of foundries and forges; chemicals; synthetic textiles; pulp and paper; cement; automobiles; home electric appliances such as televisions, air-conditioners and refrigerators; and electronic appliances and office machines including personal computers; as well as electric power generation; gas utility; the petroleum industry; and the distribution industry. These 15 industries share 73% of the total production of the manufacturing sector, and 85% of CO_2 emission from the manufacturing sector as a whole. Therefore, the study of these 15 industries would encompass practically all the major business activities in Japan.

Second, the Vision showed a way for industry to incorporate environmental concerns into each phase of the product life-cycle. Such activities used to be carried out individually at each company. MITI has successfully integrated individual efforts, and promoted voluntary action by industries to integrate environmental measures into their business activities, through the establishment of the Voluntary Plan for Environment.

Around the time the Environmental Vision of Industries was introduced, MITI also instituted a programme to support the establishment of environmental management and monitoring systems under the ISO 14000 series, for the purpose of furthering the reduction of environmental burdens through the development of creative measures in industries.

The Environmental Vision of Industries successfully clarified and showed a way for each industry to incorporate environmental concerns into business activities more comprehensively and systematically. Also, it illustrated a way to address a particular type and feature of environmental issues. In short, the Environmental Vision of Industries presented a menu of effective measures to reduce the environmental burden in 15 industries, and endeavoured to encourage individual corporations' efforts to develop and adopt such measures. This can be seen typically in the case of the iron and steel manufacturing industry. The Vision encouraged the development of high-quality steel boards, particularly high-tensile steel boards, so as to lighten the weight of commodities such as automobiles made from such steel boards, and thereby achieve higher fuel efficiency and a reduction in fuel consumption.

Third, the Vision proposed inter-industrial collaboration as a way of reducing the environmental burden at every phase of a product's life-cycle. It also identified a systematic way to promote co-operation between different industries. Examples include the increased use of recycled resources for raw materials as a way of incorporating environmental concern in the raw material procurement phase. The establishment of quality standards for recycled materials in co-operation with both manufacturers and consumers of recycled materials also promoted the use of recycled materials. Coal ashes generated from the electric power industry, for example, can be used as a raw material for cement production, and the Vision proposed to establish a JIS (Japanese International Standard) to standardise the quality of coal ashes in order to promote and secure its utilisation.

Follow-up on the Environmental Vision of Industries

The Environment Vision of Industries was proposed to address:

- The need to establish a framework for environmentally friendly business activities
- The need to develop environmentally friendly technologies
- The need to establish an environmentally friendly social system

Several measures for these have already been implemented. Regarding a framework for environmentally friendly business activities, there has been the introduction of the ISO 14000 series to establish environment management systems, and the establishing of national standards for environmental monitoring. The LCA Japan Forum has been established as a research institute to study life-cycle assessment.

In technology development measures, a programme to develop recycling technology is expected to promote overall acceptance of recycled materials in the raw-material procurement phase.

With regard to social system build-up, various relations were modified under the guidance of the Environmental Vision of Industries, such as those on waste treatment and recycling.

Furthermore, MITI plans to re-review and adjust the Environmental Vision of Industries. The review started in 1996 in view of the ongoing international negotiation for the reduction in emissions of greenhouse gases, such as CO_2, as the Third Conference of Parties of the UNFCCC (United Nations Framework Convention on Climate Change) approaches. It focuses particularly on CO_2 emission reduction and energy-saving issues. It lists measures for individual corporations, and inter-industries to reduce CO_2 emission and to improve energy savings. The establishment and review of the Environmental Vision of Industries may have stimulated the institution of the Industrial Environmental Action Plan developed among industries by the Keidanren.

The Industry and Environment Vision, instituted by MITI, has played and will continue to play an important role in showing the future direction for environmental policy-making for the Japanese government. It was instrumental in promoting voluntary actions among industries to incorporate environmental concerns into their business activities. It is MITI's intention to reinforce its efforts to promote the incorporation of environmental concern in industries, and to encourage inter-industry co-operation, focused on product life-cycle, through a comprehensive package of environmental policies.

PARTNERSHIPS BETWEEN GOVERNMENT AND BUSINESS IN ARGENTINA

Paris-based journalist Lawrence J. Speer wrote about the Argentine Campana-Zarate Project, an example of a partnership between government and industry, in *International Environment Reporter*, 5 March 1997 (published by The Bureau of National Affairs, Inc.). Key points are summarised below.

THE CAMPANA-ZARATE ENVIRONMENTAL CARE AGREEMENT IN ARGENTINA

Lawrence J. Speer

In Argentina, where environmental priorities are particularly focused on air and water pollution in the main metropolitan areas and industrial waste treatment and disposal, a new co-operative project between industrialists and local government in an industrial area north of Buenos Aires may provide the future model for self-regulation.

The Campana-Zarate (CZ) Project is a voluntary private-sector initiative to incorporate sustainable development principles into industrial policy across the region, which is home to some 180,000 people, according to Julio Garcia Velasco, Environment and Technology Director of the multinational Techint Group. (Garcia Velasco presented his project at the ECO 1997 conference in Paris.)

Techint Group, which operates a steel factory in the region, is a founding member of the CZ Project, which has 15 major industrial members, including chemicals producers and refineries. The project runs an environmental resources and planning centre, which has conducted technical studies on air and water pollution, noise measurement and potential impacts of new factory installations.

In 1993, the CZ Project took a major step towards self-regulation with the drafting of an Environmental Care Agreement, committing members to permanent impact minimisation, compliance with all environmental laws, open communications policies and co-operation with government in policy-making and control.

Today, the project members hope 'to make the self-regulatory approach a reality in Argentina', Garcia Velasco said, adding that 'implementation of ISO 14000 is a major part of this effort'.

Although the government has made moves towards simplification, CZ Project members maintain that a lack of co-ordination among the various levels of government creates problems dealing with an environmental code that remains both confusing and contradictory. The multiple compliance checks demanded by a host of regulatory agencies are costly, both in terms of paperwork and employee time, Garcia Velasco said, and could be eliminated under a self-regulatory framework. Among the points CZ Project members hope to negotiate with the Argentine government is the incorporation of a single regulatory compliance inspection for ISO 14000-certified firms.

Part 3

Tools for change

There are many new tools to help meet humanity's needs with fewer resources, and they are rapidly developing. For example, governmental policy tools in the environmental field have moved from top-down command-and-control to market incentives, and now to information-based strategies. Of course this progressive movement is not the same worldwide, nor are the categories of tools mutually exclusive. And information-based strategies are new enough for little to have been written about their effectiveness.

Similarly, management tools incorporating environmental concerns, and the newer field of financial tools (now often on the Internet), are evolving almost as fast as one can write about them. Included also herein are texts on design tools, analytical tools and technology tools, all of them evolving at top speed.

GOVERNMENTAL POLICY TOOLS

We begin our examination of governmental policy tools from the point of view of the organisation that looks at policy and performance on behalf of the industrialised world, the Organisation for Economic Co-operation and Development (OECD), located in Paris. Twenty years ago, the Environment Directorate of the OECD introduced the extremely useful tool, the '**polluter-pays principle**', into the debate. Since then the Directorate has continued to conduct research on strategies for integrating environmental concerns into the economies of the world. In working for all 29 member countries—the club of developed nations—the OECD does an analysis and evaluation of the environmental policies of each country, leading the way in its role as dispenser of rational wisdom. Policy packages of taxes/subsidy reform, regulations, voluntary agreements, 'right to know' initiatives and government purchasing are just some of the tools for change that governments are using to help achieve sustainability.

Following are excerpts from a 1997 talk in Paris by American Bill Long (OECD Director, Environment Directorate, at the time) on the evolution of tools and strategies for environmental management up to that time.

AN OVERVIEW OF TOOLS AND STRATEGIES FOR ENVIRONMENTAL MANAGEMENT

Bill Long

The tools of environmental management have evolved very rapidly over 25 years; and today policy-makers have at their disposal an extensive arsenal of strategic and technical instruments. Further, the pace of change is accelerating, for there is under way in OECD countries what amounts to a large experiment to develop a 'next generation' of management tools to cope with a new generation of environmental challenges.

To make some general observations at the outset:

- Interest and activity has shifted markedly in recent years.
- In most cases, individual policy instruments are being used in combination, and many 'hybrids' are evident.
- The choice of instruments for *common* problems is likely to vary quite markedly from country to country.

Taken together, these tools for environmental management—as applied within a market economy—have a number of interrelated objectives:

- To get the prices right (and send the right price signals to the market)—to stimulate technological innovation and diffusion
- To change behaviour
- To better inform the 'buyers' and 'sellers' of goods and services about environmental costs and benefits

Regarding the oft-quoted policy objective of changing behaviour, we might consider for a moment the question, 'Whose behaviour are we trying to change?'. There are, in fact, a range of 'actors' that environmental management tools influence—either directly or indirectly. I will simply note the high degree of attention

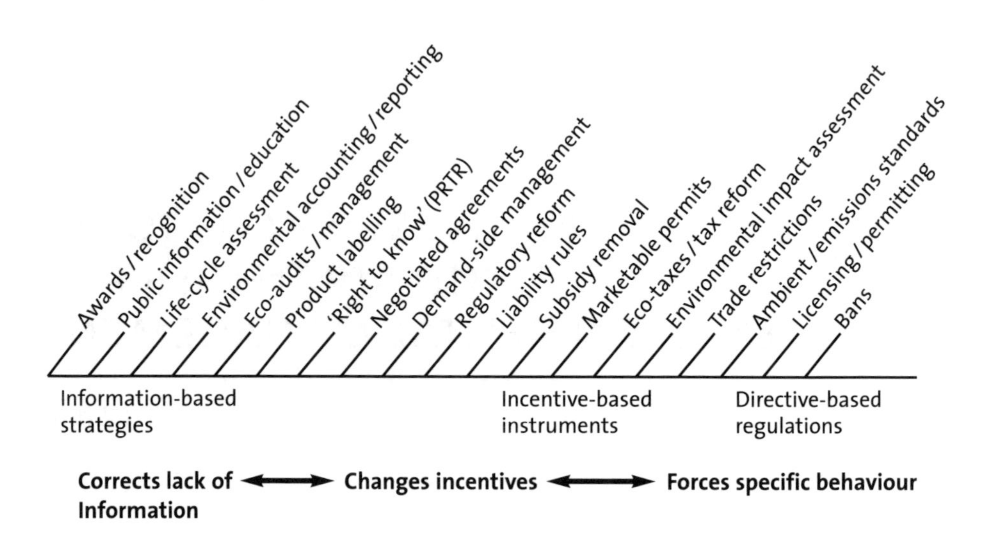

Figure 25 **Range of instruments for environmental policy**

Source: Adapted from B. Long, OECD, 1997

being paid to the so-called 'greening of government', particularly with respect to procurement policies and operation of facilities, and also to the international implications of the environmental policy tools being introduced at the domestic level.

Since the early 1970s, the evolution in the choice and design of environmental management tools has been striking. This evolution is a reflection of

- The changing nature of environmental management priorities
- The need for policy-makers to achieve greater **economic** efficiency in the pursuit of environmental objectives
- The need to avoid and eliminate **conflicts** between economic and environmental policies
- The experiences (good and bad) of government and industry in environmental management over some three decades

Several years ago, a CEO of a large multinational corporation described the changes as the '3 Ds'. He observed that, in the 1970s, the 'D' represented industry's **denial** that it was the villain, as it resisted government pressures to change; in the 1980s, the 'D' stood for **data**, with government and industry battling over whose data was better as a basis for deciding how much change was necessary and what tools to use; but in the 1990s the 'D' stands for **dialogue**, with industry and government agreeing that it is better to work together than to fight apart. Some might argue that this is an overly optimistic view of the current situation; but it does, in my view, describe quite accurately an encouraging general trend towards effective industry–government 'partnerships'.

It is important to examine some of the details of this changing panorama. For this purpose it is useful to consider what has happened over the years with environmental **regulation**—government 'command-and-control' strategies. Regulation is worth looking at for at least three reasons: (1) it was the first tool employed by OECD countries as they addressed environmental issues in the late 1960s; (2) regulation continues to form the bedrock of environmental management, and will probably do so well into the future; and (3) **reform** of government regulation across all sectors is a current preoccupation of virtually all OECD countries.

Regulation was the natural tool for policy-makers to grasp when environmental problems began to command attention in the late 1960s. This was a time of responding to crises: pesticide contamination of wildlife; highly publicised marine oil spills; 'killer' smog incidents in London and Pittsburgh; and reports of 'dying' lakes and rivers from industrial wastes. OECD governments, operating in a crisis management mentality, turned to command-and-control strategies and policy instruments that they judged to be quick and sure.

Attacking environmental pollution and resource degradation was also perceived, in legalistic terms, as a 'crime and punishment' undertaking. Today, the mentality is more inclined towards achieving an efficient and equitable **balancing** of national economic and environmental policies.

Nonetheless, for three decades, a rule-based regulatory approach to environmental management has pervaded and persisted. This is, in part, because regulatory approaches have certain intrinsic strengths—in part because each alternative to regulation has its own weaknesses; and, in part, because of government inertia. Industry support for regulation has been uneven. Some large firms have found that regulation provides good opportunities for bargaining and negotiation; and others have sensed that it can be a vehicle to increase market shares if they can get regulators to favour their approaches or technologies ('regulatory capture').

In the 1970s, first-generation environmental regulation focused on **remediation** and on correcting past 'wrongs' that posed risks to human health and were degrading the quality of air, water and land resources. Industry was perceived as the villain, with most regulations targeted at pollution and wastes emanating from fixed industrial installations.

By the early 1980s, governments were beginning to get a grip on the traditional sources of pollution—environmental 'insults' that could be seen, smelled or heard. At the same time, regulators were coming to share industry's complaints about the stifling 'cumulative impact' of the command-and-control approach. Protests were heard about both the rapid expansion in the number of regulations and their growing complexity. And, in particular, regulations were springing up in an uncoordinated fashion from independent air, water and solid waste divisions of the same national environment agencies.

By the mid-1980s, 'reform'—aimed at rationalising and improving regulatory machinery and procedures—was under way in many OECD countries. One of the goals (one that continues to receive attention) was to compress and streamline regulations in order to reduce the administrative burden and other costs to industry.

A second major goal of regulatory reform in the 1980s was to design a 'next generation' of policy tools. These were needed:

- To manage pollutants that existing legislation was just transferring from one medium to another (requiring new integrated, multimedia strategies)
- To influence industry to move away from end-of-pipe capture of pollutants in favour of reduction-at-the-source, 'preventative', manufacturing processes
- To cope with widely dispersed *non-point* sources of pollution (e.g. agricultural run-off; environmental impacts of transportation systems).

Thus, by the mid-1980s, environmental regulators were showing stronger interest in using **market-based** strategies and tools—to improve economic efficiency and environmental effectiveness, and to promote technological innovation.

The need to modify regulatory approaches, and to find new tools for new challenges, was reinforced in the second half of the 1980s as concern grew over **global-scale** environmental threats, specifically stratospheric ozone depletion and climate change. And, at the same time, environmental policy-makers were being challenged aggressively by economic interests—in government, and in industry, to take better account of the economic consequences of environmental policy measures for trade, competitiveness, investment and employment.

This began to define the international dimension of environmental policy-making, which includes a continuing quest to 'harmonise' and otherwise co-ordinate environmental policy tools at regional and global levels.

Through the decade of the 1990s, government efforts to improve the quality of environmental regulation, and to find new management tools, accelerated. This is partially due to continuing pressure from industry for regulatory relief in a period of widespread budget deficits and intensifying international competition. Industry clearly has found a more receptive audience in this climate than in earlier years.

Today, environment policy-makers in OECD governments are under heavy pressure to demonstrate—to industry, to economic counterparts in other ministries, and to the general public—that their policies are efficient and 'delivering value for money'. The pressure to find new management tools also comes from recognition that the environmental challenges of the future are likely to be much more difficult than in the past. The costs of cleaning the air and water continue to rise. Further, the international economic (and political) impacts of environmental protection measures are likely to become more consequential in this 'age of globalisation'. Bearing testimony to this are the concerns that developing countries are voicing about real or perceived economic consequences (for them) of OECD country eco-taxes, eco-labels on traded products, and emerging policies to deal with climate change.

As the decade of the 1990s comes to a close, environmental policy-making—and the tools to implement it—is in a distinctive, and dynamic, third generation.

Government efforts are moving down three tracks. First, streamlining and upgrading of the quality of existing regulations is receiving high priority. The second track involves an intensive search for **non-regulatory alternatives**—with a focus on utilising **market-based incentives, voluntary measures and expanded information about environmental risk** to bring about changes in environmentally damaging processes and behaviour. The third track involves a search for the best *mixes* of policy instruments for dealing with particular problems (i.e. command-and-control instruments used in combination with economic instruments and voluntary approaches).

Experience gained over three decades indicates that there are few common solutions to environmental problems, and certainly no 'silver bullets'. The 'correct' policy tool depends heavily on national cultures, public attitudes about environmental threats, legislative and administrative structures, and other characteristics that can vary widely from country to country. For example, why is it that Austria and Switzerland can apply the same level of 'eco-tax' to fertilisers and pesticides to decrease their use and get strikingly different results? (The answer in this case seems to be the association of an aggressive public information campaign in one of the countries.)

An important factor in the evolution of environmental management tools in the late 1990s is **the enhanced reputation that industry enjoys** as a positive force for environmental improvement. This results from increased public acceptance of industry's message that it is possible to be both 'clean and profitable'. Examples are growing of companies that have turned the necessity of improving their environmental performance into a profitable virtue, rethinking their processes and products, and coming up with solutions that have not only benefited the environment but also improved their products, sales and raw material consumption.

This has opened up an array of new opportunities for government policy-makers. Prominent among them is the possibility of entering into **voluntary agreements** with industry, with a much better likelihood of gaining public support. This is allowing regulators to shift their attention from the environmental leaders in industry (invariably the larger firms), and to focus more on stimulating and assisting improved environmental performance by medium-sized and small firms.

Up to this point I have painted a broad panoramic view of environmental policy tools, and have traced their general evolution since the late 1960s, using changing attitudes towards government regulation as the principal reference point.

I now propose to address the **alternatives and complements** to regulation, and then to conclude by referring to several specific applications of environmental policy tools that highlight current attitudes and directions.

Economic instruments

OECD surveys indicate that, among the array of market-based tools for environmental management, **taxes, charges and deposit–refund systems** are in widest use today—and their use is expanding in our member countries. Significantly, this is also occurring beyond the OECD family of nations. As one very notable example, last year Chinese authorities approached the OECD to initiate a dialogue on the 'use of market-based instruments for environmental management', and a programme of co-operation with China is now under way.

Many creative efforts to apply economic instruments are under way in OECD countries. Take, for example, OECD's newest member, South Korea. Until 1994, the sale of bottled water in Korea was illegal, as the government did not wish to alarm the public about water quality. Now there is a 20% charge for producers and importers of bottled water, with the revenues applied to ground-water and drinking water protection.

Korea has also introduced a volume-based charge for household waste. Consumers purchase plastic waste disposal bags from the grocer, and are then charged on the basis of the number of bags set out for collection. Recent data indicate that urban household waste has been reduced by 37%, and recycling increased by 40%. Increased burning of waste by households has, however, emerged as a quite serious air quality problem.

There is also a trend towards the **creation of markets**—particularly in the form of 'permits' (for pollution releases or for fish catches) which can be bought, sold and traded among various parties. Tradable permits are getting a 'hard look' as a possible tool for coping with climate change on an international level. As with other economic instruments, permits can provide strong incentives for polluters and resource users to improve technology on a continuing basis.

Deposit–refund schemes were originally applied to returnable bottles, and have proved to be quite effective. Within the past five years, the related and more expansive concept known as 'extended producer responsibility' has begun to take hold. The philosophy is that the manufacturer should bear responsibility for a broad range of products from cradle to grave—for example, by reclaiming cars, refrigerators and other appliances from consumers after the product's useful life. To cite Korea again, under a system introduced in 1995, both producers and importers of certain products pay a deposit which is refunded when the products are reclaimed. Government figures indicate that this has resulted in a 49% recovery of used tyres; a 47% recovery of batteries; a 23% recovery of lubricants; and an 8% recovery of beverage containers. This is pressuring manufacturers—in Korea as well as in other countries—to design products with longer life-spans, and/or which can be disassembled easily for re-use or recycling. At the same time, the **international trade** implications of extended producer responsibility strategies has emerged to become a matter of considerable debate and analysis.

It is noteworthy that—despite their well-publicised theoretical advantages over command-and-control approaches—the application of economic instruments for environmental management has been rather slow in developing. These advantages include delivering environmental benefits at lower cost, and providing a continuing incentive for technological innovation (i.e. their 'dynamic efficiency' function). One of the difficulties they face is the political reality that—unlike government regulation—the costs of economic instruments (especially charges and taxes) are strikingly visible to consumers and to firms. Further, industry often expresses the concern that 'eco-taxes' will merely be additive to the regulatory compliance costs they face, rather than a substitution; and that environmental taxes too often are used by government to raise new general revenue rather than to achieve environmental objectives.

This aversion to new taxes has given rise to the concept of 'green tax reform'—including major modification of national tax systems—whereby the rising costs of environmental protection (e.g. through new taxes on pollution and resource consumption) are offset by reduced taxes on labour and capital. Such 'revenue-neutral' approaches have been (and are being) studied by high-level government commissions in a number of OECD countries, notably Sweden, the Netherlands, Norway and Denmark. OECD believes that fiscal and environmental policies both should and can be made mutually reinforcing.

Current government efforts to reform markets—especially to reduce subsidies in the sectors of energy, agriculture and transport—have important implications for environmental management. For example, pressures for government regulation should lessen as elimination of production subsidies leads to reduced use of pesticides and fertilisers, and decreased cultivation of marginal lands. And the move away from subsidisation of agricultural production towards direct income support, and agri-environment payments, to farmers (to maintain the countryside and wildlife) is another implicit market-based instrument for environment management.

On the other hand, market liberalisation and regulatory reform have the potential to exacerbate and create environmental problems. For example, in the energy sector, deregulation could trigger expanded use of low-quality fossil fuels. And the implications for energy efficiency programmes, and for the expansion of renewable energy, are unclear. In light of the heavy pressures governments are under to liberalise markets by, *inter alia,* dismantling regulatory structures, the possible environmental costs of such strategies to promote economic growth need close scrutiny.

An OECD paper prepared for the June 1997 Special Session of the UN General Assembly on Environment and Development contains these conclusions about 'economic instruments' as applied in the field of environment:

- By harnessing the power of the market, economic instruments contribute to the integration of environmental concerns into economic policies by bringing market signals closer to their socially desirable levels—a necessary condition for sustainable development.

- Economic instruments have, to date, generally been introduced as supplements to regulation, with a view to collecting revenue, rather than to change the environmental behaviour of economic agents.

- There continues to be little systematic evaluation of the performance of economic instruments on environmental policy, or of policy instruments more generally.
- An important element of promoting sustainable development is ensuring that existing government policies do not encourage unsustainable behaviour. A review of existing subsidy schemes and taxes that serve as disincentives to sound environmental practices will contribute to this goal. A key challenge will be to identify clear cases for 'win–win' situations, in which both the environment and the economy can benefit from policy reform.

Voluntary approaches and negotiated agreements

Voluntary agreements between industry and government increased rapidly in the 1990s in OECD countries. Their potential for achieving environmental goals without the administrative and transaction costs of regulatory and economic instruments makes them attractive policy choices. In many instances, industry is agreeing to take measures that go beyond existing regulatory requirements, in exchange for commitments by governments to provide various types of regulatory 'relief': for example, in the form of simplified permitting processes or diminished frequency of reporting and inspections.

In some cases, the agreements are not exactly 'voluntary'. The New Zealand Environment Minister, for example, gave domestic firms three years to 'voluntarily' meet a national target for CO_2 reduction—after which he would regulate if 'voluntarism' does not work. The political acceptability of voluntary (or 'negotiated') agreements today depends on public confidence that industry is both fully committed, and able, to meet its end of the bargain. The chemical industry's unilateral 'Responsible Care' initiative launched by Canadian industry in 1985 set the stage for a 'new partnership' with government. The US Environmental Protection Agency's '33/50 Program' demonstrated in the first half of the 1990s that industry could indeed meet, and even exceed, tough negotiated targets for reducing the use of a broad array of toxic substances.

To date, voluntary agreements have mostly involved larger firms, usually the environmental 'pathfinders'. Co-operation provides corporate leadership with greater 'ownership' of environmental commitments; and also the enhanced public visibility which can come from good performance. However, many firms that are willing, in principle, to enter into voluntary arrangements are conditioning their participation on government's ability to provide a minimum base of regulation to prevent 'free-riding' by their competitors. This is another example of the trends towards the judicious use of 'mixes' of policy tools to address environmental problems.

Information-based approaches

I recall an advertising slogan played repeatedly on the radio in the US in my younger years: 'An informed consumer is our most cherished customer.' (The fact that this message was delivered on behalf of a manufacturer of fur coats probably should not be revealed to this audience.) It makes a good point, however: 'working with the grain of the market' to protect the environment clearly requires knowledgeable buyers and sellers of goods and services.

This philosophy lies at the heart of a broad spectrum of information-based tools that have emerged over the past two decades. These include eco-audits, eco-labelling, pollutant release inventories, cost–benefit analysis, and life-cycle assessment.

Public disclosure of environment-related information by industry, either voluntarily or under legislative mandate, is rapidly becoming a widely used environmental management tool. **'Community right-to-know'** programmes, designed to provide greater public access to information about the sources and nature of industrial emissions, are providing a powerful stimulus for improved environmental stewardship by industry—and in countries beyond the OECD.

The government of Indonesia and the World Bank have recently reported successful results from a pilot effort on 'public disclosure'. This involved publishing in local newspapers the names of the most egregious industrial polluters. This approach was taken when it became clear that, at present, Indonesia lacks the institutional and financial capacity to rigorously monitor and enforce compliance with its pollution control legislation.

The OECD is also supporting information-based, non-regulatory approaches, a recent example being the negotiation last year of an OECD Council Act on 'Pollutant Release and Transfer Registers' (PRTRs). This consists of a set of principles, and discrete procedures and actions, for government and industry to follow to provide citizens with expanded information about sources and potential impacts of industrial pollution. Although this was developed as a recommendation to OECD governments, several of our members are moving to place the PRTR approach on a statutory footing. Further, chemical firms in North America and Europe are beginning to use the results from PRTR programmes—which were set up to register pollution—as a new tool in production design as well as in their occupational health programmes (safer alternatives) and factory safety plans (accident prevention).

Editor's note. PRTRs have turned out to be a powerful tool indeed (see 'Internet activism', page 299). PRTRs allow the public to get involved in local pollution control and emergency response planning. Both investors and the public can compare the pollution performance of installations in different parts of their countries. For companies, PRTRs can identify operational inefficiencies that can lead to cost reductions and a competitive edge over their less eco-efficient rivals. The US system is the first and most comprehensive PRTR system, created in 1986. In Canada, a voluntary scheme, Accelerated Reduction/ Elimination of Toxics, helped to stimulate the parallel mandatory system, the 1993 National Pollutant Release Inventory. In Europe, an inventory should be up and running by 2002.

International interest has grown rapidly in recent years in **standardised 'environment management systems'** for use by industrial firms. This builds on the experience previously acquired by industry in carrying out **'eco-audits'** of plant operations. Firms continue to report that 'eco-auditing' is enabling them to both better control pollution emissions and increase profitability by conserving raw materials formerly wasted.

Particular attention is being focused at the moment on a new set of voluntary environmental management standards from the International Organisation for Standardisation (ISO). Many in industry and in government environment agencies see these standards—known as the **'ISO 14000 series'**—as a promising framework for redesigning the way that industry is regulated. While waiting for ISO 14000, European businesses have been registering their facilities under the European Eco-management and Audit Scheme (EMAS), or the British Standard on environmental management systems. Some 400 firms are registered in 1997 under EMAS, 293 headquartered in Germany.

Companies, and particularly those with big export businesses, envision important competitive advantages accruing from obtaining 'certification' under such schemes, as well as the prospect of regulatory relief and better credit rating from financial institutions. At the same time, government environment agencies are enticed by new opportunities for regulatory reform emerging from ISO 14000. One difficulty—which has generated opposition from non-governmental environmental groups—is that ISO 14000 is a *voluntary* management standard: it does not set performance targets; nor does it require public disclosure of the steps taken by firms in response to the standard, nor of the actual effects on the environment. Strong **compliance monitoring** will undoubtedly have to be part of any regulatory track based on such environmental management systems.

Information-based environmental management tools also include 'eco-labels'. The use of labelling to identify 'environmentally friendly'—or 'friendlier'—products is growing, including in developing countries. This is despite the concern registered within the trade community, and by developing countries, that such labelling creates non-tariff trade barriers. The OECD is studying how trade concerns are being—and could be—addressed in the design and conduct of eco-labelling.

Another tool that has attracted considerable attention in the last few years is **'life-cycle assessment'**. This concept was advanced, in part, by interest in eco-labelling since, 'How can one decide if a product is friendly to the environment unless one investigates its environmental impacts from the generation of the raw material inputs through the product's final disposal?'. Methodological problems in carrying out life-cycle assessments are considerable, however—as attested to by the experience of the European Commission and an array of OECD governments and various private firms. Nonetheless, it can be a powerful tool for controlling pollution at the source, even in the absence of a complete analysis. For example, the AT&T corporation has adopted a 'life-cycle' philosophy and approach even though it has encountered difficulty with rigorous life-cycle assessment, given the delays that comprehensive analysis were causing in bringing products to market. The corporation's approach at present is to assemble company experts who would operate at key points along the design, production and consumption chain of a proposed new product to discuss where the major environmental problems might crop up, and to wrestle with the question, 'What if we did it this way instead?'.

Case studies

I will conclude by citing examples of current trends and directions in the use of environmental management tools.

Ozone protection in the United States

In January 1996, the United States, the world's leading consumer of ozone-depleting chemicals, met the international deadline for phasing out production of CFCs (chlorofluorocarbons) for domestic use. This was done on schedule, and without devastating the national economy, or destroying industrial firms, as some had predicted. Today, US industry has developed alternatives for virtually all CFC applications; and the public has not been denied popular products. In case after case, firms eliminated CFCs at faster than predicted rates, at lower cost, and with greater technological improvements than anyone had foreseen.

Two innovative, market-based tools were used to accomplish this. One—initiated in 1988 by the Environment Protection Agency—involved the use of a **permit system** that controlled the production and importation of CFCs. Permits (called 'allowances') were allocated to US manufacturers and importers on an annual basis. They could be bought and sold; and the number of permits was reduced each year to squeeze down total national production and consumption. This was coupled with a **tax** on ozone-depleting chemicals, which was increased each year, raising revenue in the process and giving users a financial incentive to conserve the chemicals and to adopt alternatives.

In a recent evaluation of this experience, the Washington-based World Resources Institute drew five conclusions that are useful lessons for the design and use of environmental management tools more broadly:

- It is crucial to have management tools that can be adjusted to reflect new scientific information. (In the case of CFCs, the international community kept adjusting its phase-out targets as accumulating scientific evidence confirmed the seriousness of the problem.)
- Economic instruments can help government and industry achieve environmental goals with greater flexibility and at lower cost.
- Innovative government initiatives can remove barriers that keep industry from solving environmental problems cost-effectively. (The US Environmental Protection Agency worked closely with industry, serving as an information broker to help firms find alternatives to CFCs, and accelerating the approval process for CFC substitutes.)

- Given the opportunity, industry leaders can find ways to innovate and gain competitive advantages in response to environmental challenges.
- Initial cost projections often far exceed the actual cost of complying with environmental regulations, principally because they fail to reflect the technological innovation that environmental tools can stimulate.

Norwegian Green Tax Commission

In late 1994, a 'Green Tax Commission' was established by the Norwegian government. Comprised of both finance and environment experts, the Commission has considered the scope for new **eco-taxes** to improve the effectiveness and efficiency of Norway's approach to environmental management. It has also assessed how environmental policies might contribute to macroeconomic goals, including employment, and the potential for reducing environmentally damaging subsidies and distortionary taxes. One of the Commission's conclusions is that a broad-based 'green tax reform', with a gradual shift in taxation away from labour and towards environmentally damaging products—together with reduced subsidisation of products and activities—will help to strengthen the environment of Norway!

With respect to the question of what **policy tools** are likely to be most cost-effective in addressing Norway's priority environmental problems, the Commission concluded (not surprisingly) that greater reliance should be placed on the use of economic instruments, often in combination with regulation. What was rather surprising was the large number of times that the Commission concluded that maintaining the existing *regulation*-based approach to coping with particular environmental problems was the best approach. The Commission's conclusions were consistent with a general view across OECD countries regarding regulatory reform: the goal in the field of environment (unlike other fields, such as telecommunications) should not be seen as the *dismantling* of regulation, but rather as the improvement of the *quality* of environmental regulation.

Water pollution in the European Union

The European Commission proposed a framework directive that would make all water users pay the **full economic cost** of water—in a drive to eliminate serious water pollution and to reduce waste. A major focus is on dispersed pollution from agricultural operations, with farmers expected to pay the full cost of the degradation caused by the use of fertilisers and pesticides. The draft directive defines the 'full' cost of water as including a charge for environmental costs on top of operational and management costs, capital costs and reserves for future investment. This typifies expanding efforts in many OECD countries to include environmental damage costs in the price of goods and services.

It is also noteworthy that the European Commission believes that the directive can only be fully successful if carried out in conjunction with reform of the Common Agricultural Policy, a policy that encourages over-production and excessive use of agricultural chemicals. This illustrates both the importance of policy reform as a vital environmental management tool, and the use of different tools in combination to achieve environmental objectives most efficiently. Whether this particular directive will ever be adopted within the European Union is unclear. However, the fact that a policy instrument of this scope and content is even considered is noteworthy; and it reflects trends and thinking in environmental management well beyond the EU.

Environmental effects of transport

Controlling the environmental impacts of transport appears to many observers to be the *ultimate* environmental challenge. Many strategies and tools are being tried, with no one claiming much success to date. In a contribution prepared for the UN General Assembly Special Session on Environment and Development in June 1997, the European Conference of Ministers of Transport (ECMT; an OECD affiliate) set out some of the current thinking in the transport community environmental policy.

The ECMT paper states that ways must be found to pass on to users, as directly as possible, the environmental costs of transport services, if individual behaviour is to be changed significantly. While observing that this will require a mix of regulatory and pricing instruments, the ECMT concludes that this does not necessarily mean increased prices (since road transport is already heavily taxed in some countries). Rather, prices should be better structured and targeted to provide stronger incentives for behavioural change. The paper goes on to say that **road pricing** is the most promising instrument for internalising environmental costs over the long term; and that this should involve varying charge rates according to the time of day (higher charges for peak times) and also to the specific pollution characteristics of the vehicles. It is noteworthy in this.

Regarding the choice of instruments, the ECMT states that, although introducing market forces is likely to be the most effective approach to improving efficiency in transport, complementary regulatory interventions will be required. These regulations must, however, be designed to minimise perverse secondary effects (such as discouraging technical innovation or retarding adjustments to changes in commodity or labour markets). While regulations have proved successful in reducing the levels of some pollutants from the transport sector, they have proved less efficient at influencing fuel efficiency and CO_2 emissions; and economic instruments are likely to offer better results. One approach that the ECMT endorses is product labelling *combined* with higher taxes for certain CO_2 sources. The ECMT paper also notes the importance of **land use and development planning** as an essential part of broader efforts to reduce the environmental impact of the transport sector.

Sweden is working within the European Union to push the frontiers for bringing cleaner vehicles into the marketplace. It has proposed a new environment rating system for cars, lorries and buses which could provide the basis for differentiated road taxes and help manufacturers in marketing 'green' vehicles.

Sweden's Environmental Protection Agency claims that existing approaches—in Sweden and throughout Europe—deal with 'yesterday's problems'—such as nitrogen oxide and particulates. What is needed are policy measures to address global warming, emission of pollutants that cause asthma and allergies, and end-of-life management of vehicles. The rating system being designed would award points for factors such as emission of CO_2, emissions of specific pollutants considered hazardous to health, the efficiency and durability of a vehicle's emission control system, and the potential for recovery after scrapping. For example, some vehicles might be exempted from annual road taxes for up to five years, depending on the environmental characteristics.

Climate change management in Canada

In 1992 Canada established a multi-stakeholder 'Canadian Economic Instruments Collaborative' (EIC) to investigate the application of economic instruments to air quality issues. In a recent report, the Collaborative addressed the climate change issue. It recommended that providing economic incentives to reduce greenhouse gas emissions requires, as a first step, eliminating subsidies and other policies that serve as barriers to 'no regrets' actions, and concurrently moving towards full-cost pricing for energy. After considering specific economic instruments, particularly carbon taxes and emission trading, the EIC proposed a two-part 'hybrid' instrument, one that would combine the flexibility of tradable credits with a charge levied on CO_2 emissions from large stationary sources, as well as on the carbon content of fossil fuels used by small stationary and mobile sources, to provide a clear price signal.

Life-cycle impacts of automotive parts

The US Environmental Protection Agency (EPA) is recommending to car manufacturers that they develop methods to measure the 'life-cycle impact' of the parts they assemble into finished automobiles. According to the EPA, co-operation between parts suppliers and auto assemblers could lead to a reduction in the use of environmentally damaging materials, boost the use of recycled materials, lead to process designs that encourage recycling, and minimise environmental burdens across the life-cycle of parts.

Advocates of this approach point out that, if auto manufacturers begin to collect data related to life-cycle impacts of parts, they could exert tremendous pressure on suppliers to reduce environmental impacts. The motivating factor for manufacturers would be not only environmental responsibility and obligations, but also the idea that a less environmentally damaging product is often the most efficiently produced one. Manufacturers might also see this as a way to reduce environmental liability. The EPA argues that 'Environmental responsibility should be shared along the supply chain; final manufacturers should not have to bear the entire burden.'

At the same time, the EPA acknowledges that the concept is limited at present by the severe lack of data—on energy, solid waste and emissions—for each life-cycle stage of most products. Most data collection for environmental purposes is currently driven by regulatory reporting requirements, and does not lend itself readily to life-cycle assessment applications.

From the OECD's surveys of member countries, and case studies such as the foregoing, it is possible to identify a range of considerations that policy-makers are taking into account today as they design and select environment management strategies and tools:

- Environmental effectiveness
- Economic efficiency
- Incentive function (pollution reduction, technology innovation)
- Flexibility
- Simplicity of operation
- Cost of implementation (monitoring, licensing, enforcement)
- Integration in sectorial policies (cost internalisation, removal of policy conflicts)
- Social equity (minimisation of regressive distributional effects)
- Economic impact (prices, employment, industry profitability, economic growth)
- Trade and international competitiveness impacts
- Conformity with international agreements
- Political acceptability

While each and every policy instrument cannot be expected to score high in every category, nonetheless these considerations are all weighing heavily today in the design and selection of policy tools for environmental management. This represents a considerable step forward from the early 1970s when the approach was largely one of deciding on which regulation would control a particular environmental problem in the shortest period of time.

GOVERNMENTAL POLICY TOOLS
From command and control to governance

When it comes to a government's role in improving environmental performance, what works best as the new century begins: regulation, incentives, or governance (creating the conditions for change)? How do governments achieve improved environmental performance after 'command and control' ceases to be appropriate? The solutions depend on where the country in question is on the learning/development curve.

Vietnam, for example, created its first environmental laws in 1994 and developed an overall strategy in 2000. Russia does not have a functioning regulatory system: the country has eco-taxes, but they remain uncollected, and President Vladimir Putin abolished Russia's lone agency for environmental protection in May 2000.

China, with 8% a year growth in GDP, suffers from such severe pollution problems that even President Jiang Zemin speaks out on the subject now. A commitment to addressing environmental problems was part of China's 50-year anniversary celebration, even if it took prohibiting Beijing's largest, most polluting industries from burning coal and oil for two weeks prior to the event in order temporarily to lift the city from its usual acrid smog. Of more long-term significance in a country highly invested in its export market, ISO 14001 certification is now required for all exporters, according to Zheng Yanan, Director of the China Centre for Environmental Management Systems. So far, this requirement has substantially reduced consumption of raw materials and energy, as well as waste and emission levels of certified companies. Most significantly, perhaps, when a company is certified, others nearby are impressed and follow suit. This is like tossing a pebble in a lake and resonant rings spread outward.

The Organisation for Economic Co-operation and Development (OECD) continues to have the overview, the large perspective. Joke Waller-Hunter, who is Dutch, is Director (since 1998) of the Environment Directorate for the Paris-based OECD. She is quite direct in suggesting that, ideally, governments should abolish all subsidies that aren't cost-effective for the environment, including subsidies for energy, transport and agriculture—a total of US$500 billion a year spent by governments worldwide. Drop them all, she argues, and move as well from a tax on labour (income tax) to taxes on depletion of resources and on pollution. It is a question of getting the prices right. These kinds of proposals have more credibility, Waller-Hunter points out, when made by a country's ministry of finance rather than by its ministry of environment. She further asserts that, to date, the OECD is unconvinced of the effectiveness of voluntary agreements, whether they be unilateral (an industry-based effort), public-voluntary (developed by a public authority), or negotiated (resulting from bargains struck between the public and an industry).

WHATEVER WORKS

Based on a presentation by Arthur H. Rosenfeld

Arthur H. Rosenfeld, Senior Advisor to the Assistant Secretary for Energy Efficiency and Renewable Energy at the United States Department of Energy during the Clinton Administration, who earned his PhD in 1954 under Enrico Fermi at the University of Chicago, believes that both incentives and regulation are secondary to having the right technology. A particle physicist and author of best-selling books, Rosenfeld switched to end-use of energy in 1974. From 1975 until 1994, he was the Director of the Center for Building Science at the Lawrence Berkeley Laboratory, where he spearheaded energy-efficient developments one after the other. With the World Wide Fund for Nature (WWF), he and Assistant Energy Secretary Dr Joseph Romm created the Center for Energy and Climate Solutions, designed to help environmentally committed businesses develop and adopt innovative climate and energy solutions while gaining international recognition for their efforts.

When asked what governments should do to improve environmental performance, Art Rosenfeld says 'whatever works'. He is, for example, a firm believer in contests or 'deals', whereby energy-efficient schemes are rewarded by local governments with, for example, lower energy rates. Nevertheless, he understands the power of standards. For example, according to Rosenfeld, the energy efficiency of refrigerators in the United States has improved by 5% a year for the last 25 years because of increasing standards. The US Department of Energy, responsible for setting these increasing standards, establishes appliance standards on the basis of those proposed by the national laboratories, such as the Lawrence Berkeley Laboratory. And where do the laboratories get their ideas? They follow technological developments, and then they hold a long series of meetings, negotiations, arriving eventually at a consensus. In each case, there must be a 3–4 year payback for the manufacturer, or the labs will not forward a new, higher standard.

To demonstrate compliance to standards, appliance manufacturers must place energy-use labels on appliances for sale in the US. In the exemplary case of refrigerators, these energy labels have been an extremely effective tool. Energy saved by the improved refrigerator standards equals 45 huge power plants not built in the United States. This represents a $16 billion a year saving, which is more than the cost of nuclear energy per year in the US. Rosenfeld points out that, if finished, the Chinese Three Gorges Dam will be operational in 2016, but it will be totally outdated by then if China does not begin building energy-efficient appliances.

It is well accepted now that to increase energy efficiency, energy companies should not sell kilowatt-hours but should sell services instead. In California, Rosenfeld's group at the Lawrence Berkeley National Laboratory encouraged the government to change energy company profit rules. The proposal: every time the customers saved a dollar, they got 15 cents, which meant they were now getting a 30% return on their investment instead of 6%. As a result, the numbers of retrofits of buildings in California exploded. A building retrofit, says Rosenfeld, generally pays for itself in 2.7 years.

In Art Rosenfeld's new project, the WWF and the Center for Energy and Climate Solutions work with a select group of companies to customise progressive business plans for reducing greenhouse gas emissions. With individual companies, they explore a number of different options that could dramatically increase the efficiency of existing buildings and factories at low cost (and, in the case of buildings, possibly off-balance sheet), taking advantage of recent advances in combined heat and power (co-generation) systems, or purchasing power generated from renewable energy sources, or integrating next-generation efficiency measures into the design of new buildings, factories, and products, and/or working with their employees, customer base and supply chain to help them all take advantage of best practices for greenhouse gas mitigation.

Following are some examples of what companies can expect to accomplish, taken from Dr Romm's book, *Cool Companies: How the Best Businesses Boost Profits and Productivity by Cutting Greenhouse Gas Emissions:*

- One US Toyota plant cut total energy use by one-third from 1991 to 1996 while doubling its output with technology that helped reduce its defect rate from three per hundred to zero.

- From 1993 to 1997, DuPont's 1,450 acre Chambers Works plant in New Jersey reduced energy use per pound of product by one-third and carbon dioxide emissions per pound of product by nearly one-half. Even as production rose 9%, the total energy bill fell by more than $17 million a year.

- A California electronics manufacturer renovated one of its buildings, cutting the energy bill by 60%. The increase in daylighting and air turnover resulted in a productivity rise of more than 5% and a drop in absenteeism of 45%, bringing the payback to under a year—a return on investment of more than 100%.

- A disc-drive factory in Malaysia, Western Digital, cut energy consumption by 44% with a one-year payback while increasing air filtration requirements a thousand-fold.

- The First National Bank of Omaha is installing a fuel-cell system because it gives the most reliable electrical power available at the lowest life-cycle cost. It also cuts carbon dioxide emissions by more than one-third compared to a traditional uninterruptible power supply system.
- Dow Chemical Company's Louisiana Division began an annual contest in 1982 asking employees to find energy-saving projects that paid for themselves in less than a year. Year after year, contest winners have redesigned processes to increase productivity through pollution prevention. Savings to Dow at that one division have exceeded $150 million so far, and the search continues.
- Xerox's Palo Alto Research Center in California cut energy use in half through advanced lighting, window films, and an energy management control system.

According to Rosenfeld, through its Climate Saver Programme, the Center for Energy and Climate Solutions can help companies meet and surpass these achievements. Climate Saver business leaders will be recognised for their environmental leadership and will be associated with the WWF, an international conservation group that is at the forefront of the climate issue and the development of creative ways to address it. This association with the WWF will help Climate Saver companies position themselves as environmental leaders and integrate environmental thinking into their overall strategic planning. The Centre has assembled a team of recognised experts in key areas: energy efficiency building upgrades, co-generation, motor efficiency and industrial process improvement, renewable energy, and environmental regulations.

Meanwhile, the Energy Department, in collaboration with General Electric, has announced a major breakthrough in natural gas-powered generating plants that will result in production of electricity using 5.3% less fuel than the best previous technologies. Not only will the production of electricity be cheaper but the new technology will lead to a significant reduction in global-warming gases.

GOVERNMENTAL POLICY TOOLS

Governance (creating the conditions for change)

Lisa Lund was the United States Environmental Protection Agency Deputy Associate Administrator for the Office of Reinvention, within the Office of the Administrator when she wrote the following presentation. This means that she was the person working on adapting the EPA to meet the challenges of the new century. She is now EPA Acting Deputy Director, Office of Compliance. A trained scientist, Lisa Lund worked at the state level in environmental protection in Arizona before moving to the national level.

Lund's Project XL (eXcellence and Leadership) is a form of governance as a tool for obtaining superior environmental performance—through innovative, experimental, flexible approaches. Lund says that, once open to the new system, project participants—companies or local governments—tend to reveal all their secrets, which is a big change in environmental protection experience, where the regulating agency has in the past been seen as the enemy. In fact, the degree of openness and public participation required by Project XL might suggest that, instead of being included here, in this section on governmental tools, the idea belongs in Part 4, 'Civic action'. This is indicative of a sea change in environmental protection, where new alliances are forming.

The following extracts are from a presentation Lisa Lund made on behalf of the EPA in Paris, in 1999.

PROJECT XL

Good for the environment, good for business, good for communities

Lisa C. Lund

The Environmental Protection Agency (EPA) has been experimenting with change in order to reinvent environmental regulation. The EPA has had much success with the 'command-and-control' approach to solving historic environmental problems, but the challenges that remain today are not so suited to that approach. Problems tend to be more complex and cross statutory, media, state, regional and international boundaries. We also wish to take advantage of technological advances, and make sure our regulations are not hindering their use or effectiveness. We wish to recognise the more sophisticated stakeholders that now exist (including the state-level agencies) and to leverage their information, experience, perspectives and resources. Finally, we wish to underscore and cultivate the change in philosophy from pollution control to pollution prevention, and to highlight new awareness in the area of environmental justice.

Project XL stands for eXcellence and Leadership, and it is a series of 50 experiments that test innovative approaches to environmental management that, if successful, can be integrated into our system and lead to systematic change. XL projects must lead to better environmental outcomes, offer operational flexibility and be supported by affected stakeholders. XL is a learning laboratory, and there have been many rough patches. It is a way of acknowledging that economic growth and environmental quality are not at odds, as we used to think. Our toolkit of available legal mechanisms beyond enforcement includes permits, waivers, variances, interpretative statements, site-specific rules and deviation from existing practices and policies as allowed by statute. Project XL has taught us many things about crafting new relationships and building a system of environmental protection that will be sustainable into the 21st century.

There are eight project criteria: superior environmental results; cost savings and paperwork reduction; stakeholder support; innovation and multimedia pollution prevention; transferability; feasibility; monitoring, reporting and evaluation; and no shifting of risk burden. Project XL is open to proposals from both industry and communities. The first participants in the programme are: 3M, Intel, Berry, Hadco, the South Coast Air Quality Management District, Lucent, the Minnesota Pollution Control Agency, Merck, Anheuser-Busch, and Weyerhaeuser.

There were numerous problems at beginning of the Project:

- EPA staff were suspicious that regulatory flexibility meant that environmental protections would be rolled back.
- States felt that Project XL should be delegated to them to run. Federal-versus-state issues were thus raised.
- Environmentalists were suspicious of regulatory flexibility. They created the system we have today, and they weren't ready to set it aside.
- Industry reaction ranged from those thinking that XL was a 'free-for-all' to those who viewed it as a legitimate way of designing alternative regulatory schemes. As negotiations proceeded, many found that superior environmental performance was a difficult concept to define and implement. Some industrialists felt that it interfered with their voluntary actions, for which they wanted credit.

Initial public response to Project XL was poor, as there had been no public input. The EPA hadn't wanted to define or structure the programme too much in order to allow industry creativity in designing proposals. As a result, policy was made on an 'as needed' (crisis) basis. There was much criticism for lack of clarity. Though the EPA has engaged in more discussion with stakeholders over time, it has been difficult to overcome this initial lapse. The biggest problem was defining 'superior environmental performance', which is the whole point of the programme.

People from industry were, and still are, very concerned that, in changing the definition of compliance, participants might be held liable for traditional requirements through citizen lawsuits. Many legal issues came up around the legal standing of the agreements made for projects under Project XL. Were they legal contracts?

This is how we chose to define superior environmental performance (SEP), which is the chief goal of Project XL:

- Determine a baseline for current performance
- Compare that baseline to projected performance under the Project XL scenario
- Factor in the many subjective considerations that can make a project superior—while this may sound simple, it rarely is

Determining a baseline

If the applicant wants to eliminate lines, bring in new products, etc., determining an appropriate baseline can be complex. If they don't know or don't want to share what future products they may be producing, it puts regulatory agencies in an uncomfortable position.

Actuals versus allowables

In the Clean Air Act, facilities are granted a permitted 'potential to emit'. Companies routinely operate below allowables to avoid inadvertent exceedences. XL requires that a facility reduce actual emission levels in order to demonstrate superior performance, even if the facility was already operating below allowable level.

Voluntary controls

A company may have undertaken them before applying for XL, but these must be included in the baseline. It has been argued that this is a disincentive for environmental leaders. But EPA believes that environmentally leading companies are not doing XL to reap the incremental benefits. They are more likely seeking alternative regulatory approaches that reflect the dynamically different world that we operate in today. Still, the EPA is willing to look at the subject on a case-by-case basis.

Comparing the baseline to projected performance

A problematical aspect concerns growth or expansion. Expansion can be provided for while still demonstrating superior performance by the development of pollution-per-unit production measures. Pollution prevention projects can also deal with expansion by the design of measures for the inputs (of materials or resources) into the production process. These allow efficiency to be monitored while allowing for growth, and the EPA is very supportive of the development of these types of metric.

A compounding factor is that the EPA and the state will want these projections built into legal mechanisms that may be appropriate for the project, which in turn makes them enforceable. Many companies don't understand why the EPA needs to make XL commitments enforceable. The bottom line is that the EPA cannot give up its mandate to ensure protection of public health and the environment. So, while the EPA supports both voluntary commitments and inspirational goals, the Agency would not have the authority to enforce compliance if these were not achieved.

Factoring in subjective aspects

The Project allows for the possibility of allowing cross-media trades in granting flexibility. This is difficult as there is no accepted scientific model or tool that can compare the value and risks in a like way between media. Cross-media trades are potentially permissible if the sponsor:

- Can find a way of comparing the risks and benefits that satisfies both regulators and stakeholders

- Has the support of stakeholders
- Can demonstrate a clear benefit to their proposed cross-media trade

The EPA is leaving this issue of cross-media trades open by not explicitly naming an accepted methodology by which trades will be judged. This leaves the decision as to whether the concept is worth the investment with the potential applicant. Science has not been able to come to any good conclusion to date about comparative risk and benefit models, so it is not fair to expect XL to resolve the science on this in a hurry. What the EPA hoped for was that, by allowing the potential for cross-media trades, XL would inspire holistic analysis of facilities that would clearly prioritise risks. XL can be a vehicle to deal more quickly and rationally with the highest-priority problems, while allowing flexibility in areas posing smaller risks. The Weyerhaeuser project is a good example. They proposed cutting their bleach plant effluent—the largest concern at pulp and paper mills—in exchange for flexibility in how they achieve emission reductions.

Public participation is the most challenging component of Project XL. Early on in the programme, the EPA determined that the 'one-size-fits-all' approach to stakeholder involvement would not be suitable. The applicants—as managers of the XL process—can tailor the stakeholder involvement process as they wish. But there was confusion, and they had to re-engineer. The re-engineering process led to changes in the actual negotiation process itself as well as development of a tool: 'Project XL Stakeholder Involvement: Guide for Sponsors and Stakeholders'. The state agencies are pivotal stakeholders and have veto power over projects.

The programme has faced perceptions that a project applicant could 'orchestrate' stakeholder support and that the EPA therefore needed to better define the parameters of stakeholder involvement. The project applicant, not the EPA, is responsible for initiating and maintaining the stakeholder involvement process. The EPA does provide facilitation assistance to project applicants for initiating stakeholder processes.

Project XL should be seen as a problem-solving tool. If certain regulations are proving to be hindrances to good business practices, then this should be viewed as an opportunity to design a new approach. Some kinds of flexibility offered include: allowing process changes without prior review or permit modifications by operating under an emissions cap; using pollution prevention to achieve emission reductions instead of costly control equipment; using high-quality environmental management systems as the basis for consolidated permits; designating a multi-use development with mass transit capacity as a transportation control measure to promote smart growth; paperwork reductions and administrative cost savings due to consolidated reporting and/or alternative monitoring schemes, etc.

While many think-tanks and other groups have discussed the provision of regulatory flexibility in exchange for better performance, XL has been the first to actually do it. One should not underestimate how hard it is for all participants to move these projects forward. The process has been re-engineered with the help of all participants several times. Now there is a five-step process with an associated time-frame, as well as clear roles and responsibilities:

1. Pre-proposal—informal discussion of an applicant's concept
2. Proposal development—with the EPA plus the state in question
3. Selection in hopes of incorporating the new approach more broadly
4. Negotiations—with EPA, the state, the applicant, plus local stakeholders and local governments
5. Implementation

Environmental benefits

XL projects have resulted in reductions in emissions and discharges in air, water and waste. In one year and counting seven projects only, there have been:

- 10,422 tons pollutants reduced (nitrogen oxides [NO_x], sulphur dioxide [SO_2], particulate matter, carbon monoxide [CO])
- 1,326 tons volatile organic compounds (VOCs) reduced
- 22 tons hazardous air pollutants (HAPs) reduced
- Reductions of 2.19 lb per air-dried metric ton of biological oxygen demand in effluent
- 1,284 tons of solid waste recycled
- 415 tons of non-hazardous waste recycled
- Reductions in water and energy use (non-quantified to date)

Benefits to participants

- Financial gains
- Efficiency benefits
- Industry leadership, improved standing, reputation among consumers, hand in shaping future regulation, etc.
- Better community and stakeholder relations
- Improved relationships with regulators at all levels

Some specific examples of changes include:

- Intel won a competitive advantage in the quick-to-market semiconductor industry by avoiding millions of dollars' worth of production delays through the elimination of 30–50 permit reviews a year.
- Weyerhaeuser achieved an estimated administrative savings of $176,000 during the first year of operation under XL. The company is now saving

$200,000 a year by recovering lime muds and re-using this solid waste. It expects to avoid $10 million in future capital spending.

The extensive interactions of community and facility representatives that take place during the course of an XL project may help forge real and informed trust with the local community. Involvement with environmental organisations has been beneficial to the companies in the long run in terms of better mutual understanding. Industry projects that have traditionally lacked experience with convening and managing a site-specific stakeholder process were able to build new relationships.

So what about the transferability? Since the overall purpose of XL is to lead to system change, is any of that happening? In the first Project XL Annual Report, the EPA looked at rules and regulations, permitting, information reporting and management, stakeholder involvement, enforcement and compliance assurance, and Agency culture change. There have been changes for the better in all.

- For ongoing information and evaluation of Project XL, see www.epa.gov/ProjectXL.

GOVERNMENTAL POLICY TOOLS

Eco-taxes

On the questions of environmental financing and taxes, France has made some interesting moves, creating fiscal mechanisms for financing environmental agencies engaged in specific tasks. For example, the French Agency for Environment and Energy Conservation (ADEME) is financed partly by a tax on waste emission, and certain water districts function in the same way: there is a fiscal penalty levied on industrial emissions in order to apply pressure to reduce them. In the case of the ADEME, industry's continued production of certain industrial wastes leads to funding for research or activities seeking to mediate their production: the polluter-pays principle. Such earmarked taxes function well when the collection and use of such funds is constantly monitored. Indeed, how are French earmarked taxes working now after 15 years' operation? Is the idea exportable? Following are excerpts from a presentation made in Paris in 1997 by Jacques Vernier, then President of the ADEME.

France has developed proposals for further eco-taxes on pesticide run-offs, toxic wastes, and eventually on the country's largest energy users as a way of encouraging energy efficiency. Prepared jointly by the Environment, Economy and Industry Ministries, the proposals are a significant example of an 'integrated' approach to environmental problems as well as of taxation policy drafted to achieve both environmental and economic goals.

In 1999, the German Parliament's approval of increased eco-taxes on gasoline and electricity means energy taxes rise annually through 2004.

TAXES EARMARKED FOR ENVIRONMENTAL PROTECTION

The French experience

Jacques Vernier

Environmental taxes have become unbeatable instruments for environmental politics, unbeatable in the way that they can raise money for environmental investments, unbeatable in the way that they marry the 'polluter-pays principle' to the rule that he who doesn't pollute or who invests in order to avoid polluting is given a fiscal reward. Yet such taxes are not implemented randomly but via a large, varied and integrated structure of regulation, regional or departmental urban planning, and financial and technical tools. Many different public bodies are involved: the state, regional groups, water agencies, plus the Agency for Environment and Energy Conservation (ADEME). The agencies are at the heart of the collection and dispersal of tax monies regarding water, air, waste, oil and airport noise.

Use of taxes earmarked for environmental protection began in France in the 1960s, first in the water sector, with the creation of legislation in 1964 and then applied to water pollution beginning in 1969. In the 1980s, the practice was extended to air pollution, with a tax on atmospheric pollution established in 1985. Household waste became subject to a tax for treatment and disposal in 1992, and, since 1995, industrial waste taxes contribute to a special fund for the rehabilitation of polluted sites. There are other instruments, such as those used by Eco-Emballages SA, in which fees paid by businesses cover the costs of separated waste collection and sorting of waste packaging. Other earmarked taxes relate to airport noise and development that maintains natural open space.

The growth in earmarked taxes is part of, and a reinforcement of, environmental policy incorporating national policy, European policy and international commitments made by France.

The need for earmarked taxes can be found in ambitious management and pollution-prevention objectives. Most funding for national spending on the environment—by the Ministry of Environment, or by other ministries or agencies working with the Ministry of Environment—comes from taxes on air and water pollution, and on waste.

To take some examples: in 1997, the budget of the Environment Ministry was approximately 2 billion francs; the monies from taxes on air and noise pollution, oils and waste managed by the ADEME was 1.1 billion francs. The budgets of Water Agencies reach 50.9 million francs (from user fees, from industrial and domestic pollution, and from repayment of infrastructure loans).

Earmarked taxes are often criticised in the name of neutral, any-use fiscality, because such taxes are assessed by decree on specific sites and source points and are often maintained at the same level for several years. It would of course be best if inter-ministerial negotiations took the public interest to heart when it came to environmental protection, but national budget deficits can undermine the best efforts of the Environment Ministry when it comes to any-use monies. Indeed, what one sees is a general reduction of general-fund monies as earmarked monies increase. The ADEME can attest to this in regard to its budget, and it is thus that energy conservation, even at this crucial moment, has become the poor relative of other environmental policies.

With regard to the annual versus pluri-annual budget, the advantage for environmental agencies lies in the visibility and need for possible long-term funding when it comes to the considerable investments required. Short-term funding would simply not suffice. And, furthermore, there is some democratic control over even long-term taxes: the tax on atmospheric pollution was reconsidered in 1990 and 1995 before being reapplied long-term. Other democratic controls include water-basin commissions or commissions charged with managing taxes on air pollution or waste, along with regional authorities and non-governmental organisations.

So it is difficult to imagine today what system might replace earmarked taxes as a means of reinforcing environmental protection, and, as a matter of fact, most stakeholders approve of the system.

Small earmarked taxes linking the polluter-pays principle and the principle of aiding those who invest to avoid polluting are part of a complex system of regulation and incentives. These taxes, 35 francs per ton of household waste in January 1997, twice that for industrial waste, or 180 francs per ton of, for example nitrates, exist in place of the less-accepted eco-taxes which might more accurately reflect environmental costs.

Realistically, there is no agreement on the real economic costs of certain activities on the environment or on human health. Further, the introduction of genuine eco-taxes is still a politically, socially and economically unacceptable solution. Issues of competitiveness in the absence of international harmonisation of standards, costs and prices are problematical. Many arguments exist in opposition to economic environmental theory as put forward by international institutions such as the OECD. Yet we are even now on the verge of having an eco-tax on CO_2 emissions. Earmarked taxes are an evolution in the right direction.

While not everyone is in agreement about the respective merits of various economic incentives versus harsh regulations, most policy analysts agree to the need to send a signal to the market about the costs of environmental degradation. Thus, modes of production and consumption which cause less damage to the environment are encouraged, and other more destructive modes are discouraged. Fiscal instruments available to accomplish such ends go beyond earmarked taxes to include tax deductions or lower taxes on processes that respect the environment. The simple fact of making the environment a factor in taxation plays a major role. Investors are made aware of the best ways to avoid pollution or to treat emissions, discharges and general waste, or else they pay according to the damage caused. Logically, revenues from environmental taxation ought to end up as part of the state's general funds, thus avoiding a specific administration for earmarked taxes.

But things are not so simple. The facts show that, over the long term, the tax on petroleum products which raises the price of fuel to the consumer in France has had a beneficial effect on technological choices. This doesn't mean there have been no problems, such as when this kind of tax goes counter to good environmental sense, such as when prices go up on unleaded gas and down on diesel.

On the other hand, if we were to suddenly introduce genuine eco-taxes, internalised within costs, even if we were to satisfactorily settle questions of international harmonisation, of neutrality, the situation would still be brutally abrupt, leading to the need for a new collective mobilisation to refinance in accordance with environmental objectives. The goal of environmental protection might not be reached, and no level of taxation could immediately ameliorate real environmental damage. With environmental demands simply a budget item in the general budget, we are back to arbitrating among various public needs.

Earmarked taxes that fund specific environmental objectives appear to have several advantages:

- They signal the public interest in pollution prevention; there is both a clear interest in dealing with certain kinds of pollution and an availability of funds.
- They apply the polluter-pays principle at an acceptable level.
- They favour economic and environmental efficiency with a tax that is low and because those who invest to reduce pollution are aided.
- Because they are graduated according to the level of pollution or use (as in the tax on waste), they are clear and allow the user to adapt and anticipate.
- They allow for financing of intervention programmes and encourage innovation.

MANAGEMENT TOOLS

Strategic environmental management

The integration of strategic environmental management systems into business operations represents a new phase in environmental protection and a move away from the 'command-and-control' approach, when regulation by governments at all levels was the standard approach to pollution prevention.

The following excerpts are from papers written by two experts from the French Agency for Environment and Energy Conservation (ADEME), François Demarcq and Valérie Martin. Demarcq is presently Director General of the ADEME. Earlier, he was Secretary General of Soccram, a private company specialising in energy services, where, among other assignments, he specialised in developing urban co-generation systems. He also worked at the European Bank for Reconstruction and Development creating a multilateral fund to finance improved security in some of the more dangerous nuclear plants in central and eastern Europe. Valérie Martin is an economist at the ADEME responsible for research in environmental management, in the socioeconomy of waste, and in eco-industries.

ENVIRONMENTAL POLICY OF BUSINESSES

Evolution and future vision

François Demarcq and Valérie Martin

Integration of environmental concerns into business practices is, like production methods and quality control, a challenge that businesses today must accept if they are to survive in the marketplace. Businesses are now subject to growing external and internal pressures to integrate environmental considerations into their strategies: increasingly aware citizens, elected officials and administrations, client pressures, the development of international standards, the evolution of regulations, the growing interest from the financial sector. Environmental awareness can also be seen as part of the social role of business with the new emphasis on environmental quality as part of working conditions.

The 1987 Brundtland Report, *Our Common Future* (WCED 1997), warned of the dangers implied by environmental deterioration and identified poverty as its major cause, and, thus, environmental protection took on a new international role. Analysing the politics of each country, the report showed that environmental concerns must be part of development in order that development be sustainable. Sustainable development was defined as development that meets the needs of the present without compromising the capacity to meet the needs of future generations. The report became a basis of the 1992 Rio Summit and recognition of the need for a better balance between economic and environmental variables. Legislative activities of the 1990s are evidence of a transition from a system based on repairing damage to the environment to a system based on preventing damage. New recognition of environmental questions has also come from the business sector: the International Chamber of Commerce Declaration, eco-labels, eco-audit regulation, the ISO 14000 series, etc.

Up until the present moment, most national environmental policies have been based on the 'command-and-control' approach. Progressively, public officials are engaged in promoting new structures that encourage rather than penalise, which make environmental concerns a factor of development rather than a constraint factor. Among the changes envisioned is the involvement of various economic players in both the definition and application of environmental policy. Further, there is now international talk of 'Factor 4' on the horizon, the use of four times less energy and resources for the same product or service, which is why governments are now developing new strategies: evolving regulation, increased research and development, diffusion of best available technologies, multilateral accords, etc. For businesses, new strategies include certification to ISO 14000 standards and full implementation of strategic environmental management systems.

Toward a new kind of environmental management

With environmental management spreading rapidly throughout the world, the issue becomes how to reconcile industrial strategy and environmental policy. Environmental concerns are entering into industrial strategies. The questions are these: What is at stake when a business adopts environmental management? What is the role of various environmental management tools? How do these tools compete with one another? Can a business get along without them?

Various business motivations are:

- To assure better internal organisation, identification of tasks and actors
- For managers, to develop a better knowledge of the factory and its work flow
- To increase environmental awareness and to anticipate future regulation
- To be recognised as a good-citizen business; to be socially responsible
- To integrate environmental soundness as a driving factor equal to quality

Within a community, motivations are:

- To better integrate the business into the community
- To create confidence and goodwill for the business *vis-à-vis* the local power structure
- To improve communication with the media (and not get caught short)

In the marketplace, motivations are:

- To meet the demands of large clients
- To beat the competition
- To assure the final buyer of the environmental quality of the product and the business

Faced with this new challenge, strategies adopted by various businesses are multiple. A. Louppe and A. Rocaboy (1994) distinguish between five possible strategies: a hostile response (all ecological concerns are by nature uneconomical), a defensive response (the environmental factor is considered a threat), an accepting attitude (the environmental factor is a legitimate social concern but not the responsibility of business), a co-operative response (a willingness to be involved in environmental objectives), and a proactive response (managers integrate environmental soundness into their quality objectives). The last is still a rare response.

Origin and history of the ISO 14000 standards

ISO 14000 standards have become increasingly accessible as a result of all the work done on them by various experts. The International Organisation for Standardisation (ISO) developed the ISO 14000 standards through the work of the Technical Committee 207 whose objective was to provide interested organisations with a common approach to environmental management.

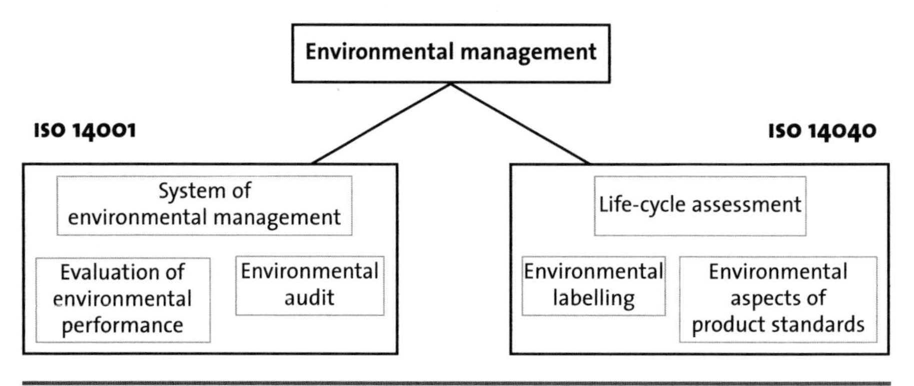

Figure 26 **Organisation of Technical Committee 207**

ISO 14001 standards

'Technological warning system', 'tool for observing competitors or for protecting one's products', 'window onto technical evolutions'—these descriptions are all attributed to the standards. Published in 1996, the ISO 14000 standards 'directives for a management system' define the principles and the recommendations for starting up an environmental management system. The ISO 14001 'specifications and instructions for use' define the requirements a system ought to satisfy. They become the means by which a third party can certify that a business has demonstrated its capacity to control environmental impacts. All businesses, whatever their size or their sector, can obtain an ISO 14001 certification. It can apply to a particular site or to the whole business. Local governments can also apply such standards to waste treatment plants, to industrial zones or to the maintenance of open spaces. The philosophy of ISO 14001 is based on its voluntary nature, and on its definition of objectives and targets. Such systems do not replace regulation, but can aid and stimulate industry in developing beyond regulation.

The use of ISO 14001 standards can be seen in four ways:

1. Internal functioning, interrelationship between divisions concerned with environmental questions
2. Creation of inter-site leadership (competition between sites)
3. External assistance, as in winning contracts where certification is a requirement
4. The certification itself as a source of competitive advantage.

Environmental management is occurring in both developed and developing countries, even though at first it was perceived as an obstacle in trade within certain developing countries. In fact, it is facilitating trade. At the end of 1998, there were a total of 7,966 ISO 14001-certified sites in the world, with 1,542 in Japan, 1,100 in Germany, 950 in the UK, 463 in Korea, 398 in Taiwan, 210 in France, 210 in the United States, 120 in Belgium and 40 in Hong Kong.

The European Eco-management and Audit Scheme (EMAS)

The European eco-audit regulations were original in their legalistic structure and in their operation. After various systems were developed throughout the European Union (British BS 7750 standards, French X 30-200 standards, Irish IS 310 standards, Spanish UNE 77-801 standards, etc.), the European Council adopted in June 1993 a system of regulation allowing for voluntary participation on the part of businesses. This system has been in effect since April 1995. Its objective was to encourage industrialists to adopt site-based environmental management systems and to communicate about them with the public. The system is entirely voluntary. At present, EMAS concerns only industrial production sites, with plans to extend it to transportation, tourism, banking and insurance. The specific objectives of EMAS are multiple: to promote continuous improvement in environmental performance through establishing policies, programmes and management systems; to establish a systematic, periodic and objective evaluation; and to furnish adequate information about the subject to the public.

At the end of 1998, 2,360 businesses were certified EMAS, with 1,795 in Germany, 152 in Austria, 64 in the UK, 21 in France and 8 in Italy. Most large German companies already have an environmental management system and adhere to EMAS. EMAS is popular in Germany and the UK because of tight relations between industrialists and the administrative organisation, leading to the relaxation of regulation as well as infrequent unexpected inspections; and because the technology is transferred readily to small and medium-sized businesses. In France, the administrative system is cumbersome, and the industrial community is unconvinced of the system's value.

Regarding the cost of putting an environmental management system in place for an installation of more than 100 persons, an ADEME-sponsored study by Paul de Backer shows that the average cost is 1,550K French francs (with half for internal costs—the salary costs for those responsible for the system), with annual operating costs going between 720K and 1,450K francs.

The development of environmental management tools beyond certification

In 1990, following the reflections that led to the elaboration of the National Plan for the Environment—the Ministry for the Environment—the ADEME wished to set up a partnership with the economic sector. The Environment Enterprise Plan is the tool of this policy. The originality of the partnership is that it concerns all companies whatever their status, activity sector or size, as well as their functions: supply, production, distribution, finance, human resources, research and development, organisation and marketing. The Environment Enterprise Plan is a tool aimed at the creation of strategic business policy as well as information and communication.

The Environment Enterprise Plan is:

- A tool for structuring the environmental policy of a company, which can define, grade and quantify the objectives as well as elaborate an action plan
- A methodological and conceptual frame to conceive and initiate an environmental policy

The ADEME has the mission to adapt the Environment Enterprise Plan to small enterprises (SMEs [small and medium-sized enterprises]/SMIs [small and medium-sized industries]). To adapt it to small and medium-sized businesses required adaptations to make it more accessible; the objective of ADEME is to help all economic actors anticipate better the evolution of regulations, to aid their decisions and to ameliorate the efficiency of their environmental management. Further, the ADEME is now working on adapting these systems to the needs of local municipalities, to aid them in developing an Agenda 21.

Conclusion

Environment appears for some companies as a strategic opportunity. Environmental questions can be transformed into new commercial opportunities and present one with a rare opportunity to distinguish oneself from competitors. Environmental management can be a means for a company to position its products and to demonstrate 'eco-citizenship' towards its employees, clients and shareholders. Environmental management is a genuine technical and organisational innovation, particularly because of the major economic issue that a better integration of environment into public and private strategies represents. However, in order to fix permanently the place of environmental questioning within companies, it is important that companies develop an integrated strategy in order to gather all involved partners together around a more positive and proactive vision of environment. Also, the set-up of an environmental management system should facilitate access to financial markets. The main goal of this specific system is the optimisation of global and environmental results. To achieve this, a system must be periodically re-evaluated and revised. The existence of such a system can offer certain guarantees to investors and insurers. The question of outside certification is problematical. Some companies have created excellent internal audits. The purpose of the system is to have a state-of-the-art standard recognised by all. It is generally voluntary and shows the user's desire to meet certain standards.

But aren't environmental standards really a means of creating hidden trade barriers or tariffs? In fact, the proliferation of national, regional and international standards, even those based on public demand, reinforces obstacles to free trade. But environmental regulations will have an increasingly important role in future markets and will thus effect competitiveness. Environmental performance will take its place within business alongside human resources; the two are equal partners in sustainable development.

MANAGEMENT TOOLS

Strategic environmental management in evolution

Ira Feldman is Senior Counsel of Tetra Tech/greentrack™ (based in Washington, DC), and Vice-Chair of the Environmental Performance Committee of the US TAG (Technical Advisory Group) for the ISO 14000 series. Mr Feldman is also Co-Chair of the Legal Issues Forum of the International Organisation for Standardisation (ISO), and was a member of the Environmental Management Task Force of the President's Council on Sustainable Development. He was formerly Special Counsel at Environmental Protection Agency (EPA) Headquarters, where he directed the Environmental Leadership Program and led the revision of the EPA Audit Policy.

In the following 1999 paper, Feldman argues from the point of view of the consultant that the gradual, international move toward strategic environmental management systems reflects the new, less adversarial approach to environmental protection as well as an actual culture change away from regulatory 'command and control'.

A NEW PLAYING FIELD?

Ira Feldman

Emerging regulatory and non-regulatory trends constitute a shift from the compliance- and liability-driven mode to the strategic and the sustainable. Strategic environmental management reflects an overall *culture change* to a more proactive, less adversarial concept of environmental management. The emergence of strategic environmental management will not only affect the range of services needed from the law firm/consulting firm perspective, but it will also cause a gradual evolution in the role of the regulator and will probably change the job description of the environmental health and safety (EH&S) staff within an organisation. Today's environmental manager is no longer limited to issues of regulatory compliance; the interests of a broad range of stakeholders have created a more complex dynamic.

The next 'quantum leap' in environmental management will be the movement towards *sustainable business practices.* Over the last 25 years the US has made significant progress through the system of 'command-and-control' regulation. Most good corporate environmental citizens are satisfied with a legitimate effort to comply fully with laws and regulations, but stakeholder pressures demand more. The ISO 14000 environmental management system approach provides a framework to make the leap from a compliance mode to an integrated environmental management programme.

Understanding the 'implementation equation'

Every business organisation needs to understand the four distinctly different drivers that underpin the need to move towards the strategic and the sustainable: (1) competitiveness; (2) internal efficiency; (3) regulatory flexibility; and (4) external financial considerations.

The example of the Japanese consumer electronics industry (which has completed its implementation of ISO 14000, a process begun at the suggestion of the Japanese Trade Ministry), illustrates how the environmental management system (EMS) standard has generated **global competitiveness**. Not only has environmental management become a competitiveness issue between consumer electronics manufacturers based in various countries, it has also created a 'domino effect' in the Pacific rim and ASEAN countries as related to the supplier chain of the Japanese consumer electronic manufacturers.

It is not necessary for the top company in the supplier chain to have committed to implementing ISO 14000 in order to affect second- and third-tier suppliers. As supplier chains are being downsized, having an ISO 14000 certificate or showing other evidence of an environmental management system may not be determinative, but it is rapidly becoming part of the expectation in certain industries.

Similarly, the competitiveness driver appears in the public sector through procurement guidelines. One example in the United States is the requirement by the Department of Energy (DoE) for its contractor-operated facilities to show evidence of environmental management systems as each management contract comes up for renewal. This has led to four DoE facilities implementing environmental management systems in conformance with ISO 14000.

The **internal efficiency** driver should be relevant to those organisations that have had a good experience in implementing an ISO 9000 quality management system (QMS). An ISO 14000 EMS, with many of the same components and terminology as ISO 9000, can be integrated and aligned with the QMS. If implemented correctly, the result should be cost savings tied to internal efficiency. Further, implementing an environmental management system, through a cross-functional team approach, allows for better integration of the environmental health and safety programme into the overall management system of an organisation. This consideration is critical in those organisations where the EH&S function is viewed as a costly resource drain.

The **regulatory flexibility** driver for strategic environmental management is critical, since ISO 14000 environmental management triggers far greater public policy implications than ISO 9000 quality management ever did. The role of voluntary excellence pilot programmes such as Project XL, the Environmental Leadership Program and the Merit Partnership, need to be noted here, as well as the 'big-picture' policy efforts of the Enterprise for Environment, the National Environmental Policy Institute, the National Academy of Public Administrators and the Aspen Institute. Regulatory flexibility is not speculative—it is available today through innovative proactive approaches and participation in these regulatory flexibility programmes.

The **external financial** driver suggests that preferred rates from lenders and reduced premiums from insurers will follow, when the financial sector appreciates the relationship between EMS principles and risk management methodologies. The United Nations Environment Programme (UNEP) has already fielded an international programme under which insurers are asked to sign a memorandum of understanding (MOU) linking EMS to risk management. Over 70 insurance companies have signed onto the MOU, but at present only a handful of them are North American companies. As domestic financial sector interest increases, the implementation of a strategic environmental management programme will be properly viewed as an effort to reduce the risk profile of the organisation. Such efforts will be seen by investors and Wall Street as an indicator of good overall management practices.

Putting the four drivers together—competitiveness, internal efficiency, regulatory flexibility and external financial—an organisation can create its own **implementation equation.** This analysis is equally applicable to a small business, a large manufacturing facility, a service organisation, a municipality or a county, and can be used as a framework for determining how an organisation may benefit from a proactive strategic environmental management approach.

A global context

The United States is no longer the unquestioned world leader in environmental protection. We may still be the leader in *environmental compliance* issues, and we may have the best in terms of innovative *environmental technology*; however, it is just as clear that the Japanese, the Germans, the Dutch and the Scandinavians have moved ahead of us in utilising certain *environmental management* approaches.

Given transactions such as the DaimlerChrysler deal with multinational implications, it is clear that in a global economy US businesses will need to track what the rest of the world thinks about environmental management tools.

The domestic solution must begin by exploring new alternatives to the 'command-and-control' box, and by embracing the sustainable business practices that will define the path forward and the emergence of strategic environmental management. If indeed the future of environmental management is now, then certain key trends—domestic and international—are worth watching, as summarised below.

Domestic

- EPA Office of Reinvention activities, especially the redesign of the flagship XL programme. An 'Innovations Taskforce Report' has just been issued in August 1999.
- Integration of the voluntary excellence programmes based on EMS core components, such as the ELP (Environmental Leadership Program of the EPA), Merit Partnership and Star Track
- Financial-sector involvement, through the National Advisory Council for Environmental Policy and Technology (NACEPT; established by the EPA) capital markets committee or continuation of the Aspen Institute dialogue
- Sector-specific (versus media-specific) advances, especially for industries that need to use strategic environmental management as a competitive tool following deregulation, e.g. the utilities sector
- Sustainable communities. Municipalities, counties and other entities will be basing their programmes and activities around formalised sustainability indicators and formalised sustainability plans.

- Follow-on activities to the now-defunct President's Council on Sustainable Development, specifically the environmental management task force, which drafted and recommended a new framework for environmental regulation
- New directions from the Administration, such as the 'livability initiative' housed in the Council on Environmental Quality (a Presidential advisory group).

International

- Activities of the United Nations on sustainable development. The environmental programmes in UNEP and the Commission on Sustainable Development, as well as other agencies have contributed to advances in sustainable business practices.
- Alternative regulatory frameworks. Governments and multilateral institutions are experimenting and moving beyond 'command-and-control' programmes, e.g. with covenants in the Netherlands, integrated pollution control in the UK, greening the gross domestic product in Germany, and voluntary corporate initiatives in Japan.
- Sustainability as a business concept. The initiatives of the World Business Council on Sustainable Development and the European Greentable are going to have spillover effects, beginning with the Global Reporting Initiative (GRI).
- International accounting guidance. An expert group within the United Nations Conference on Trade and Development (UNCTAD) has been looking at environmental accounting methodology and disclosure for three years. A final draft is imminent.

MANAGEMENT TOOLS

Strategic environmental management in evolution

David Monsma is Senior Program Manager for the Sustainable Commerce Program at the American NGO, Business for Social Responsibility. He was Co-ordinator of the Environmental Management Task Force for the US President's Council on Sustainable Development (PCSD) and has participated in all recent US environmental management reform dialogues. The PCSD was a federal advisory committee to the President formed in 1996 and comprised of leaders from government, business, environmental and native American organisations. Unfortunately, the PCSD was retired in 1999 as being politically unpopular, along with public-policy use of the phrase 'sustainable development'. Both of these developments are indicative of a backlash against the vanguard environmental movement in the US.

The term 'livability', encompassing the more tangible and narrower goals of redevelopment and controlled urban sprawl, has replaced 'sustainable development' among some American planning-oriented environmentalists. Another such term is 'smart growth'. There is now a Smart Growth Network that works toward 'development that serves economy, community and environment'. Their website[11] says:

> Growth and development directly affect environmental quality. Many communities are now asking questions about the relationship between the number of commuter trips and regional air quality, the impact of development on habitat loss, and the consequences of impervious pavement on water quality . . . Smart Growth takes advantages of these locational decisions and site designs to minimize development's impact on the environment.

The following paper by David Monsma grew out of his participation in ECO 1999 in Paris.

11 www.smartgrowth.org

ENVIRONMENTAL MANAGEMENT IN THE GLOBAL ECONOMY

David Monsma

Sustainable development, then and now

In North America, as in Europe, Asia and Africa, 'sustainable development' is often met with pedantic scepticism. Regrettably, the term 'sustainable development' was at times used by world banks and funding institutions to palm off Western growth attitudes and infrastructures to less developed nations in need of capital flows and development assistance.[12] Public and private financiers sometimes exploited the commercial potential of unmarketed human and natural resources in developing nations through lending instruments intended to urge economic growth and instill the profit motive regardless of the impact.[13]

One unfortunate result of such policies in developing countries was the application of a 'use-it-or-lose-it' development model. Under the pattern of growth, the informal or vernacular economy was 'adjusted' and natural resources were put to use in marketed production, or sold outright ('rented') by the state to multinational and foreign interests. For instance, the economic worth of rainforests was determined by cattle grazing and the fossil fuels beneath, rather than through valuation of the informal uses intrinsic to the extant ecosystem and culture.

Support for sustainable development, particularly from US environmental groups, has been sparse mainly because these actors traditionally have resisted domestic measures to promote trade and technological innovation and diffusion. Their argument has been that globalisation efforts such as NAFTA (North Ameri-

12 See de Romana 1989, discussing alternatives to mainstream economicism through a reinterpretation of the multi-dimensional, systemic and transdisciplinary analysis of the cultural and environmental impacts of mainstream economic organisation and practice; and Orton 1994: 13-19, comparing, for example, the 'limits to growth' arguments with the advent of the sustainable development justification for economic growth and 'market-based measures' for protecting the environment.

13 See e.g. Daly 1996: 'The focus exchange value in the macroeconomic circular flow also abstracts from use value and any idea of purpose other than maximization of the circular flow of exchange value.'

can Free Trade Agreement) or GATT (General Agreement on Tariffs and Trade) promote greater economic and environmental disparity. Such concerns, while true in some respects, ignore the obvious fact that we are irrevocably woven into the very fabric of the global market economy. The question of how globalisation and technology will result in new, unforeseen problems is far more relevant than accepting a level of environmental protection functionally equivalent to economic protectionism. Free trade at any cost is not a politically acceptable outcome if environmental and economic justice count as truly democratic principles, but neither is intransigence based on an incapacity for change and an embittered political adherence to conventional and negative oppositionalism.

Notwithstanding the early commandeering of sustainable development for wholly free-market purposes, the notion may come to occupy a more bona fide position in the environmental management policies for both consuming nations and the developing world. Internationally, within the rich array of cultural perspectives on the environment, a prevalent theme arises among very different environmental managers wrestling with very different environmental conditions and economic choices. Corporations such as Interface and Novo Nordisk, municipalities as diverse as Paris and Portland, and transitioning economies such as those of Vietnam, China and Thailand are all beginning to see the multiple advantages and value of sustainable development. And nowhere is this recognition more obvious than in the potential of more adaptive, performance-based approaches in environmental management.

Strategic environmental management

The US pollution control system[14] has successfully and persistently reduced emissions and ambient concentrations of pollutants over the past 30 years. Yet, where it was once considered a model for other countries to follow, it perhaps unwittingly impedes the emergence of a more global and sustainable environmental management framework. However, many are beginning to agree that the current system, which is an agglomeration of dozens of unaligned statutes passed at different times with different regulatory philosophies, is ill suited for meeting the future, and that the rate of return on environmental protection is diminishing as the economy changes and grows.[15]

Among the countries that struggle the most to construct new environmental management programmes from the ground up, there is increasing awareness that a form of environmental management more strategic than the media-specific

14 'Pollution control system' refers to the set of nine major US environmental laws administered by the US Environmental Protection Agency. For a complete definition, see Davies and Mazurek 1998: 2-5.

15 The supposition is that existing environmental protection requirements are the lowest accepted common denominator for defining environmental outcomes and do not necessarily translate into increasing environmental improvement, especially on the scale that we are likely to encounter and need over the next 20 or 30 years. See e.g. PCSD 1996: ch. 2, 26-55, 1999: ch. 3, 34-55; Hausker 1999a; Aspen Institute 1999.

model pioneered by the US is a necessary and constructive part of advancing environmentally sustainable growth. Environmental professionals in the US have also articulated such a vision. The 1999 report of the President's Council on Sustainable Development (PCSD 1999), a White House advisory committee,[16] identified several attributes of a new environmental management framework and recommended critical steps that could move the existing system towards one that is more sustainable. These included: developing a new environmental management approach organised for improving performance; utilising environmental management systems, and third-party certification and auditing; along with better aligning environmental management with the economy.

Although there is no agreement on a universal set of protocols for a performance-based environmental management system globally, it is increasingly understood that most developing countries are unlikely to adopt large-scale regulatory frameworks such as those devised by US and Europe during the 1960s and 1970s. The cost and impracticality of implementing an entire complex of environmental requirements foreordains that it is not going to happen. Rather, the strategic use of environmental management systems (EMSs), life-cycle assessment (LCA), clean development mechanisms (CDMs) and standardised environmental performance reporting are far more likely to be adopted and quickly implemented if only because it makes eminent sense to 'leapfrog' toward a more comprehensive and self-verifying system. Moreover, the potential and logic of linking environmental performance to business performance and financial assessments at both micro- and macro-economic levels make EMSs attractive to governments whose resources dictate using cost-effective systems that promote continual improvement, provide transparency, and that can be verified (and certified) by third parties.

Insular trading systems for a discrete number and type of pollutants notwithstanding (i.e. climate change carbon markets), another more driving need exists for implementing a global environmental management framework that promotes the strategic use of EMSs and performance standards. Although the reasons differ between developing and consuming nations, the need for integral systems that aim to promote environmental performance and continuous improvement are very similar if not the same. We share the same global economy, the same transnational corporations, and fundamentally the same basic ecology. We also share

16 The PCSD was a federal advisory committee to the President formed in 1996 and comprised of leaders from government, business, environmental and native American organisations. It concluded its work in June 1999 after producing its second chartered report. Among other things, the PCSD Charter asked the council to 'Advise the President on the next steps in building the new environmental management system of the 21st century by reviewing current environmental management reforms (including Project XL and other innovations), further developing a vision of innovative environmental management that fosters sustainable development (environment, economy and equity), and recommending policy improvements and additional opportunities to advance sustainable development.' Revised Charter, PCSD, 25 April 1997; and Executive Order No. 12852, 19 July 1993; Further Amendment to Executive Order No. 12852, as amended 30 June 1997.

the need for environmental management techniques that adapt to rapidly shifting corporate restructuring and perpetual industrial migration. One obvious result is that the increasingly dynamic global economy is outstripping the capacity of the statutory environmental protection system to effectively identify, prevent and control pollution.

The global economy

Economic change in the 1990s has at least three dominant implications for environmental management: (1) the composition of what is made and emitted has changed; (2) the organisational structure and location of industries and companies has changed; and (3) the rate of production, reorganisation and relocation has accelerated. Moreover, there is every indication that this trend will continue. The corporate entity is no longer as fixed or predictable as it may have once been.

In the past 20 years the composition of economic activity in the United States and Europe has shifted away from traditional or monolithic manufacturing such as steel-making, towards service and information-based sectors. More relevant are the environmental implications of industrial realignment, organisational change, and the increasing importance of time-to-market concerns of a growing number of industries including the automobile and micro-electronic industries and their numerous vendors. Additionally, modern financial markets have created a state of constant corporate restructuring in debt and equity, acquisition, merger and short-term strategic partnerships. This makes it harder and harder for environmental managers everywhere to identify, prevent and control pollution according to what is being released, who generates it and where it is produced. The case holds true not only for relatively new, dynamic industries such as semiconductors, but traditional industries such as chemical and automobile manufacturing.

Another effect of restructuring is that it allows companies to 'unbundle' or vertically disintegrate and locate manufacturing and assembly far from corporate headquarters—in other states and countries. Whereas design, manufacture and assembly once took place under a single factory roof in Silicon Valley or Dallas, a microchip engineered and sold by a Texas firm may now be manufactured by a competitor in Dresden, Ireland or Taiwan, assembled in Singapore and sold in Sydney or Tokyo. Restructuring also allows companies to condense further the time between research and development and high-volume production. Such changes make it harder to identify and to assess current and future environmental effects under systems designed 10–20 years ago (Mazurek 1999a).

The European experience

In response, European Union (EU) member countries during the late 1980s and 1990s have shown a greater willingness than the United States to experiment with voluntary and regulatory approaches to improve environmental management and environmental quality. EU approaches include voluntary environmental management certification schemes such as EMAS (Eco-management and Audit Scheme) and mandatory efforts such as the Integrated Pollution Prevention and Control Directive to integrate individual air, water, waste and toxic permitting requirements. Moreover, at a gathering of European and Asian environmental managers in Paris in 1999, discussion focused on how to build on the use of environmental management systems such as ISO 14001 to design more sustainable, performance-oriented approaches from scratch in places such as Vietnam, Thailand and China.[17] If participation is taken as popularity, consider that there presently are over 300 voluntary environmental agreements in EU member countries and roughly 40 administered by the US Environmental Protection Agency (EPA) at the federal level in the United States (OECD 1998b).

One lesson to be drawn from the European experience with respect to voluntary agreements and to radical policy change is that the most effective, efficient programmes possess legislative backing and strong government commitment to achieve programme goals. The most notable example is that of the Netherlands. The dramatic change in approach to environmental management in the Netherlands was seen as 'necessary and inescapable' (Hersh 1996). And, in 1989, the Dutch National Institute for Public Health and the Environment concluded that, even with the full application of existing end-of-pipe technologies, it was impossible to prevent further environmental degradation and create conditions for sustainable development in the Netherlands (Hersh 1996). A more recent EU communication concluded that 'environmental agreements can offer more cost-effective solutions when implementing environmental objectives and can bring about effective measures in advance of and in supplement to legislation' (Commission of the European Communities 1996).

Ironically, perhaps, part of the European willingness to experiment with new approaches is driven by the same economic rationale used in the United States to avoid novel or voluntary approaches and preserve the uniformity of the status quo. The EU, which must undertake to normalise environmental standards by and between its member states, is beginning to look to the development of voluntary agreements and mass customisation based on the use of environmental performance standards and environmental management systems as one way to get there. In the United States, resistance to large-scale change in the environmental management framework is reinforced, in part, by the view that ambitious innovation may be in direct conflict with the federal, uniform system of laws.

US reform movement

While the major US environmental statutes and regulatory requirements give us target outcomes based on the use of best available or achievable control technology, they do not necessarily give us adequate information on environmental performance or actual environmental conditions or impacts. For instance, com-

17 ECO 1999, International Congress, Paris (7–9 June 1999).

pliance with command-and-control requirements imposed at the time of a permit or a rule gives us a baseline and a range of possible emission levels (e.g. from actual to permissible). But it does not necessarily give us adequate information on environmental performance or a good indication of actual environmental conditions or impacts. The result is static, or not necessarily improving, environmental performance.

Thus the US regulatory system, by its very nature, has become costly, inflexible, lengthy and litigious. The current US regulatory system: (1) is designed to achieve a limited objective of compliance; (2) does not readily yield a positive environmental outcome or gain; (3) does not accommodate the now prevailing public desire for disclosure and transparency; and (4) is seriously out of date with the economic realities, business structures and business practices of today.

In the US, consensus on the need and desire to create a 'bold, new alternative "performance-based" environmental management system' (PCSD 1996: 34) was formed by and between major stakeholders as early as 1996, and reiterated in one process or another every year thereafter.[18] Whereas most agree that a performance-based system is necessary for the future, few agree on the specifics. Here, the design of a new performance track[19]—one developed and operated in parallel with the existing regulatory system, utilising emerging environmental management techniques and complementary policy developments—is close at hand, but needs reinvigorated support to be applied on a greater scale and with more confidence.

The challenge of preventing and controlling pollution at a time of great change has not been lost on the US EPA. During the late 1980s and throughout the 1990s the Agency has developed and implemented at least 30 non-regulatory experiments designed to complement, integrate or reinvent the pollution control system (Mazurek 1999b). Such efforts appear increasingly popular in the regulated community. According to EPA estimates, the number of participants in the Agency's voluntary initiatives will have grown from roughly 400 in 1991 to 13,055 by 2000 (EPA 1998). The EPA's first major voluntary programme, 33/50, was designed to promote pollution prevention by creating performance goals and encouraging a voluntary reduction of emissions of 17 target chemicals by 50%. Concluded in 1995, firms exceeded the goal a year in advance. In contrast, the Common Sense Initiative and Project XL use multi-stakeholder negotiation to forge agreements that give participants flexibility to select the most cost-effective abatement methods, provided that they reduce as much, if not more than, pollution control laws require.

A closer look reveals that, in the US, voluntary approaches primarily serve as complements to compliance with existing regulations, whereas their role in

18 PCSD 1999: 113-19; Hausker 1999a: Appendix B-3; and Enterprise for the Environment 1998 (Chapter 4 lays out recommendations on the increasing use of performance-based approaches and adopting EMSs).

19 Aspen Institute 1999 outlines particular policy issues that must be addressed in offering operational flexibility in exchange for superior environmental performance and stakeholder involvement; see also Metzenbaum 1998: 71-73; Feldman 1997: 11-15.

Europe is often to serve a desirable alternative to regulation. In the US, the upkeep of the pollution control and environmental management system is used to justify the incrementalism or gradualism towards reform. These imperfectly aligned environmental laws also retard progress by pitting the federal system against innovation. At the state level this results in a distracting and tiresome conflict over federalism (i.e. some states misapply innovation by attempting to reduce or weaken the stringency of environmental requirements without substituting standards as good or better than federal law requires). Paradoxically, most regulatory experimentation in the US is restrained by the federal system since federal environmental laws provide very little systematic variance or experimentation leading to consequential modification.[20]

Conventional wisdom also imposes a rigid political adherence to the status quo—sticking to a solid but inflexible system of environmental controls which may be improved over time, although not modified greatly (and certainly not customised). In the steady-state view, tinkering with hard-earned environmental standards is naïve and potentially dangerous. Unfortunately, adversarial political circumstances, and the heavy reliance on compliance with the federal system of laws, leads away from voluntary approaches and a wider acceptance of environmental management systems in the US By contrast, the EU and its member states have used unification as an opportunity to experiment with the existing pollution control system and improve on the piecemeal approach of media-specific programmes (i.e. non-integrated air, water, toxics and soil requirements).

The New World

In the EU, wider experimentation and adoption of new systems is the best way to align environmental progress, and good ideas in one member state have greater potential application in another. However, around the world, environmental professionals operate in a veritable environmental-protection vacuum when compared with the US and industrialised Europe. Globally, in areas where the new economy has little or no environmental regulatory foundation, the diffusion and adoption of novel systems is critical and will ultimately define the future of environmental protection worldwide. One leading example is the use of environmental management systems in organisations as diverse as the Chinese government, which has an institute devoted to the development of EMS practices, and IBM, which recently certified its facilities under ISO 14001 worldwide.

There is already a wealth of data and knowledge that could inform practical efforts to implement an even more strategic use of environmental management systems worldwide. As an initial step, for instance, a common environmental management framework could employ the building blocks that have been

20 Many of the environmental statutes provide some kind of variance, although the EPA has not produced, and industry has not demanded, a comprehensive variance programme. In this sense, reinvention has served in an ad hoc way to stretch the elasticity of the current system, sometimes employing site-specific rule-making to avoid legal infirmity.

demonstrated in ISO 14001 and EMAS. The environmental community, corporations and governments can all play a role in developing such a framework, perhaps by evaluating and sharing models of EMS-'plus', adding the attributes and improvement goals currently lacking in ISO and EMSs in general. An EMS-*plus* approach would include environmental performance standards and goals, standardised core data collection and disclosure, third-party verification, and other elements now being discussed in the environmental management movement everywhere.

Conclusion

The increasing worldwide demand for lifestyles like those of the consuming nations, and the economic growth and environmental pressures that will accompany it, suggest that paradigm change, rather than incrementalism, is necessary. Economic growth in the next 20 years will not resemble the economic growth that was occurring when the pollution control system in the US and Europe was adopted in the 1960s and 1970s. Corporations do not look and do not behave now as they did then. This fact, and the determinant realities of the equity markets, are enough to indicate that environmental professionals must adapt their knowledge and skills for identifying, preventing and controlling environmental impacts in ways that anticipate this growth, that are informed by the changeover of capital, and that can match the new economy. A more globally constituted framework of interactive environmental and information management systems is the only way that our economies will fairly compete and still complement one another's ability to protect and restore ecological balance.

Although we may lack political agreement, we have the technology. We have the ability to create a more global performance-based environmental management framework that incorporates comparable and verifiable environmental standards and information systems (both voluntary and mandated). This opportunity is driven by a growing understanding of what environment sustainability means for both consuming and developing nations. If we ignore that possibility of meeting our needs now, we certainly put at risk the ability of future generations to meet their own needs. Therefore, as environmental professionals, we must work even harder to close the gap between older regulatory systems and new environmental management systems, between environmental protection methods used by consuming nations and developing ones, between inflexible but strong environmental laws and voluntary agreements—between self-limiting parochial interests of one country and the ecological wellbeing of all. There is no utopian remedy or perfected end state, but more universal approaches can and should be developed as we agree to use each other's best ideas. Still and all, 'nothing changes without enthusiasm' (Voltaire).

DESIGN TOOLS

Eco-conception

The ADEME (the French Agency for Environment and Energy Conservation) is the French environmental regulator, but it is also a research and analysis organisation. It is thus possible to measure France's progress on the integration of environmental considerations into industrial practices by looking at what the ADEME is promoting. Why is it that in 1999 the ADEME is just getting around to advancing the idea of eco-conception, when life-cycle thinking has been around for over a decade and surged early in the 1990s? Why is the agency just now moving to introduce such thinking into the national industrial fabric?

The ADEME now seeks to convert the largest possible number of stakeholders, nothing less than the whole French manufacturing sector, small and intermediate-sized and large businesses included, most of which have remained untouched by life-cycle thinking, simply because they weren't operating in the sectors most affected by regulatory or competitive crises.

What is most remarkable about the following remarks by Pierre Radanne, President of the ADEME, is the immense work and time required to transform isolated and remarkable innovations into the national norm. It was first necessary to identify and validate the best methods, the best actors, to convince various governmental administrations and agencies of the best ideas so they could study the tools, create the systems, develop the communication, etc. And, now, business leaders large and small are being told and beginning to recognise that they can spontaneously be part of the action, that they are free to originate the action if they will.

Pierre Radanne has been President of the ADEME since early 1998. Before that, he was Assistant Director of the office of the French Minister of Land-Use Planning and Environment. Radanne also founded a European institute for evaluating energy and environmental strategies (INESTENE), and served his time in the World Bank as well. Following are excerpts from a presentation he made in Paris in 1999.

ECO-CONCEPTION
Driver of environmental management and competitiveness

Pierre Radanne

Today, we, like many national environmental agencies, seek to encourage rather than penalise, and thus we are moving toward voluntary measures. Most enterprises, and we are particularly interested in the small and medium-sized businesses now, have at least heard of the site-based approach to environmental management, whether it be eco-auditing or ISO 14001 certification. A product-based approach, on the other hand, takes us outside the company through an examination of all phases of a product's life-cycle, both before and after the manufacturing process. This approach, while still rare, is the most likely to involve the final consumer in the environmental protection process.

Eco-conception is the reduction from the very beginning of all future environmental costs generated by a product. Eco-conception can stimulate pollution prevention from both the supply side and the demand side. Eco-conception is, at least at the present time, completely voluntary, but it can be encouraged and facilitated by national environmental policy. Eco-conception can succeed in a free-market economy because it leaves innovation and creativity in the hands of industry. Further, such product-based innovation is the natural complement to site-based efforts, as it allows a manufacturer to speak about the latter through the former, through the actual product.

One of the essential characteristics of eco-conception is that each manufacturer must make it their own. There is no one-size-fits-all design solution. Still marginal in France and elsewhere, eco-conception has the potential for an enormous contribution to the integration of environmental concerns into industry. The OECD, for example, is studying eco-conception as a means of accomplishing 'Factor 10', the improvement by a factor of ten in the use of energy and primary resources, and in the reduction of pollution and waste. With 220 kg of household waste per year per person in 1960, growing to 434 kg in 1995, without counting increases in industrial and agricultural wastes, it is clear that curative methods have no effect. It has become essential to decouple economic growth from an increase in material goods and waste. Prevention is the very essence of eco-conception. Prevention provides a much better handle for controlling environmental costs, and it often turns out to be more economical in other ways as well.

Eco-conception can lead to small steps, one after the other, or to giant technical and commercial advances. It can be useful to both small and large enterprises. It is even now increasingly a part of international project competitions and of specifications provided to suppliers.

Present and future perspectives

The integration of environmental constraints at the design level first manifested itself through the blacklisting of substances that should not be used in products, a listing that occurred primarily because the negative impacts of such substances had been pointed out in the media. This totally reactionary approach is obviously limited, and it ignores the potential for creative proaction by product designers when they are trained and have access to life-cycle assessment techniques. So far, the most noticeable changes in design have occurred, of course, in packaging, partly because it is so visible and appreciated by the public.

There have also been advances in eco-conception as a result of the progression from products to services. Whereas photocopy machines are now leased rather than bought, with maintenance, repair and even replacement built into the deal (all of which have contributed to design changes), other sectors are now following suit. In the office furniture sector, leasing with a contract for maintenance and replacement has led to design changes of the leased goods leading to increased durability and easy replacement of parts rather than the whole. Taken to the extreme, this substitution of services for products will lead eventually to dematerialisation (such as centralised voicemail systems instead of telephone answering machines). This is no doubt the way of the future.

In France, change often comes through working groups assembling representatives of all actors—in the case of eco-conception, this would include industry, associations, academics and those in government. One such process led to the creation of EIME™ (Environmental Information and Management Explorer), an ecodesign software tool collectively created by the Ecobilan Group, with Alcatel, Schneider Electric, Legrand, Thomson Multimedia and IBM, and financed also by the ADEME.

EIME™ identifies, at conception, the potential weaknesses of a product with regard to environmental impact, and can assist the product designer in material choices and manufacturing procedures. It can also identify a strategy for continual improvement while, at the same time, making it possible for a manufacturer to fully integrate environmental concerns into product development. Through training offered by the EIME™ licenser, non-specialists can become adept in its use.

At the national level, in an effort to promote the eco-conception of products, the ADEME launched a call for projects (up to 100,000 francs each) involving the development of such techniques particularly within small and medium-sized companies. With a budget of 3 million francs, the ADEME is helping qualifying companies become part of the eco-conception revolution (29 projects were selected, and completed by mid-2001).

DESIGN TOOLS

Design for environment

Within five years of its founding, Ecobilan was the world leader in matters of applied life-cycle assessment and was a privileged observer of the development of environmental policy in the large industrial groups. Ecobilan also functioned as a proselytiser for the techniques of life-cycle assessment as applied to industrial strategy. LCA involves accounting for all consumption and emissions during the life-cycle of a product or of several alternative products fulfilling the same function, from the extraction of primary materials through the treatment of wastes. Today, life-cycle assessment is ISO-standardised, but its uses and forms constantly evolve.

It is clear, though, that sound product conception is not just a question of software. As constraints increase, industrial conception processes pose particularly demanding management problems: how to co-ordinate into concurrent engineering the growing number of experts with possibly contradictory points of view? How to steer the process towards a technical compromise that is innovative yet goal-oriented in terms of quality, cost and design deadlines? How to efficiently articulate research objectives and solutions and the actual product-conception phase?

A consortium of electronic goods manufacturers funded a design for environment (DfE) project over the last three years, and in this framework, Ecobilan developed a DfE software tool (called EIME™) for Alcatel, Alstom, IBM, Legrand, Schneider Electric and Thomson Multimedia. There is now a growing interest among car manufacturers in design for environment. Integrated eco-conception of products is the subject of a three-year joint research project Ecobilan/Centre de Gestion Scientifique de l'École des Mines de Paris in co-operation with an electronic-sector group. Further, the Fédération de la Plasturgie has launched research on the development of an adaptation of design for environment methodology applied to their sector.

The following was prepared by the Ecobilan group in 1999, before it became part of PricewaterhouseCoopers: Rémi Coulon was General Manager of Ecobalance, Washington, DC, Pascale Jean, PhD, was Marketing Manager of Ecobilan, Paris, and Hélène Lelièvre was a Project Manager of Ecobilan, Paris. They all joined the Sustainability Department of PricewaterhouseCoopers in 2000.

BEYOND LIFE-CYCLE ASSESSMENT
An integrative approach to design for environment

Rémi Coulon, Pascale Jean and Hélène Lelièvre

What is DfE?

The basic idea of DfE is to put environment into practice as early as possible in the design of a product. It uses a variety of methods, all of them showing some usefulness, ranging from non-computerised, qualitative approaches to strictly quantitative methods, where the use of software tools is common.

For example, eco-innovation brainstorming sessions can be very valuable. Similarly, non-software methods include checklists, design guidelines, or the use of lists of targeted substances. More quantitative methods tend to rely on life-cycle assessment or inventories. Whatever the method, life-cycle thinking is practised, which means that all the life-cycle steps of the product are taken into consideration.

Environmental considerations are best integrated into a company's decisions early and upstream in the design process. *How* should these considerations be included? The experience with the electronic industry shows that three axes have to be worked on simultaneously:

- **Environment.** The definition of evaluation methods and environmental metrics is needed. Indicators must be defined specific to the products and processes of the industry under study. They can be derived from classical LCA indicators, but their relevance has to be verified, in order to make the approach light and efficient. Rules related to environment must be inventoried, including environmental regulations, external industry standards and rules internal to the company.

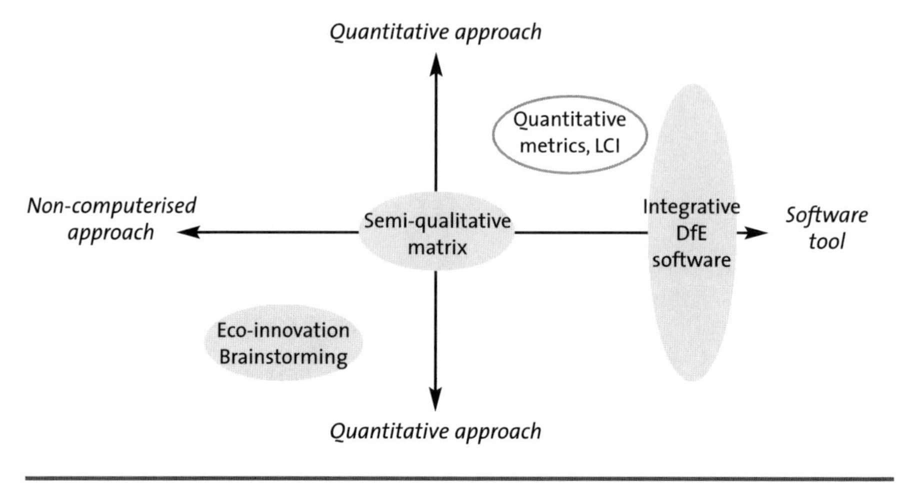

Figure 27 **A family of tools and methods**

- **Organisation.** Defining environmental indicators is not enough. The way they should be used must be defined as well. This is a key step in the DfE process: the design process must be thoroughly analysed, the stakeholders identified and information flows tracked. Then the task is to feed this process with DfE information, with the lowest disturbance and the maximum efficiency. The best solution will probably differ from one company to another, since the best solution depends on the information exchange structure of the company, as well as on its project management scheme.
- **Tools.** Finally, working on the environmental and the organisational side does not guarantee the success of the DfE process: everything learned and invented in those fields must be translated into user-friendly tools, which can give the right person the right information at the right time. These tools can take the form of checklists, guidelines, and, in the most comprehensive approaches, the use of software tools becomes necessary.

Requirements for a successful DfE approach, as derived from the electronic-sector experience, are:

1. Make designers environmentally aware and bring them value, without requesting from them a significant training effort. Designers are a key point in any DfE approach, since they are the ones who finally design the product. Designers must integrate environment as a full dimension of their task. However, the tools handed over to them must be simple and easy to use by non-environmental experts.
2. Design the DfE tool as a communication tool that can be used by designers, but also management people, marketing staff, environmental

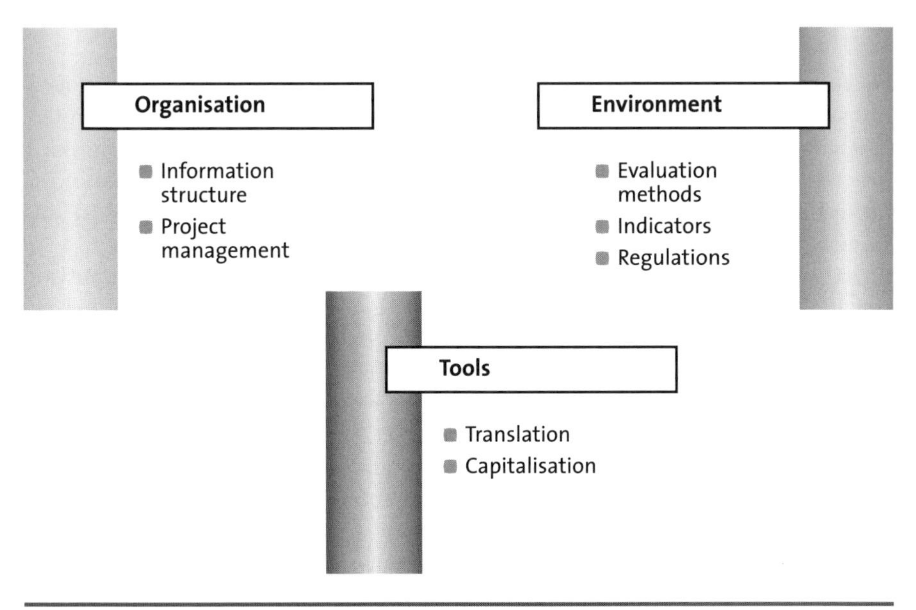

Figure 28 **Three axes for an efficient DfE approach**

experts and any staff involved in the decision-making process regarding product design. As explained above, all the stakeholders concerned with the design process must be involved, which makes a DfE tool a communication tool *first*. Since stakeholders have different interests and concerns, the tool must be customisable to meet each and every need.

Basic content of a DfE process

Let us imagine for a few minutes that we are designers, not knowing much about environment, but willing to design a 'green product'. What would we expect from a DfE tool?

1. We would like the tool to **validate the compliance** of the product with regulations, internal rules and clients' requirements.
2. We would request from the tool a quick evaluation of alternative designs, for **comparison**, based on generic data, i.e. data already gathered in the tool.
3. For this purpose, we would expect to have a limited number of relevant **indicators**. These indicators could be both the translation of LCA results into understandable indicators such as 'greenhouse effect potential' or 'water toxicity', or additional indicators, such as the recycling rate, the number of different materials in the product and the weight of hazardous substances entering the product.

Positioning DfE versus LCA

For those of us familiar with LCA, the development of DfE might pose some questions in terms of positioning DfE versus LCA. Whatever the DfE methods adopted, life-cycle thinking should remain the framework, because life-cycle thinking prevents one pollution from being transferred upstream or downstream in the life-cycle, or from one environmental compartment to the next one. In this framework, DfE goes both further and less far than LCA.

DfE is a pragmatic approach. As such, it takes short cuts: when accurate and specific data is not available, then generic data is used for DfE purposes. This can be done in LCA too, even in the framework of ISO 14040, but the uncertainty on the results has to be estimated, whereas in a DfE approach one takes the risk of being wrong. Such short cuts mean that the results of a DfE comparison are probably less precise than those of a classical LCA. As time goes by, LCA practice should feed the DfE tool with more specific and accurate data. A mature LCA practice should also help qualify the relevancy for a given DfE tool, and help draw boundaries for use of the tool.

Further, DfE covers more than the classical flows recorded in a life-cycle inventory. For example, the number of links difficult to dismantle is considered in a DfE approach, whereas it is not part of a LCA. Similarly, the compliance with regulations, though important for the company, has nothing to do with LCA, and is included in a DfE approach.

LCA is a defined quantitative science-based tool, whereas DfE is a wide pragmatic approach. They both rely on life-cycle thinking, and LCA feeds DfE with data and relevant indicators, whereas DfE helps translate LCA into day-to-day decisions, by combining it with other relevant factors, including qualitative information.

The DfE experience in the electronics industry

Five major electronic product suppliers, Alcatel, Alstom, IBM, Legrand, Schneider Electric and Thomson Multimedia, supported the development of a DfE tool, which began in 1997. The current version of the tool, available on the market under the EIME™ name, is version 1.4. It is used in limited areas for product development among those companies.

Structure of the tool

The tool was designed as a communication tool. Different people with different concerns can find the information they are looking for. There is a central database, with an Oracle structure, and a client–server architecture. There are two sets of interfaces:

- One for the non-expert (in particular designers): very simple, where only what is strictly needed is shown, and where the complexity is hidden

- One for the environmental expert: at this level, the complexity of the environmental evaluation is available, and can be managed. This level controls what is seen at the 'non-expert' level. As such, this tool is a means of transmission and diffusion of environmental strategy to the operational people.

Content of the tool

The designer has only three different screens:

- One for the description of the product content
- One for the description of the product packaging, distribution, use and end-of-life
- One for the results

Figure 29 shows the screen that allows for the description of the product content. A new design is easily built in the upper-right-hand-corner window by click-and-drag of items selected from the existing database, available on the lower right-hand corner of the screen. This database covers components, materials, processes, coatings, and also links. This is particularly important for modelling the dismantling phase.

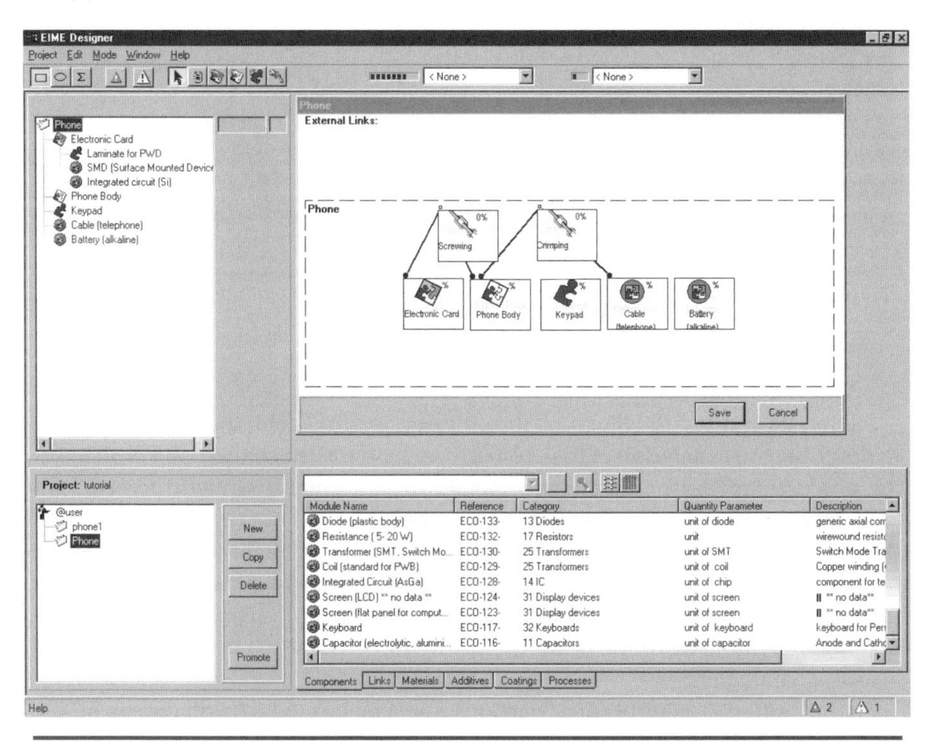

Figure 29 **EIME™ designer screen for the description of the product content**

Different levels of detail are achievable in the definition of a design scenario: a component can be selected as such in the database, or its content can be entered by the designer, if it is more relevant and if the data is available. The description relies on a nested structure: a single box, describing a component, can hide a complete description of sub-components, parts and links. Of course, any component described in a previous design exercise can be re-used by copy and paste to save time, and interesting design options can be shared among the different users of the tool.

Figure 30 shows an example of results from a simulation. Two design options for a phone are compared. Option 2 is chosen as the reference. It could be an actual design option or a chosen target, and option 1 is compared to option 2 through a limited number of indicators. Here, 11 axes have been chosen, and a spider-web graph displays the relative performance of option 1. It shows that option 2 is preferable, in particular for an indicator such as 'ozone depletion' (OD), for example.

Indicators	Short
Raw Material Depletion	RMD
Energy Depletion	EDTST
Water Depletion	WD
Global Warming	GW
Ozone Depletion	OD
Air Toxicity	AT
Photochemical Ozone Creation	POC
Air Acidification	AA
Water Toxicity	WT
Water Eutrophication	WE
Hazardous Waste Production	HWP

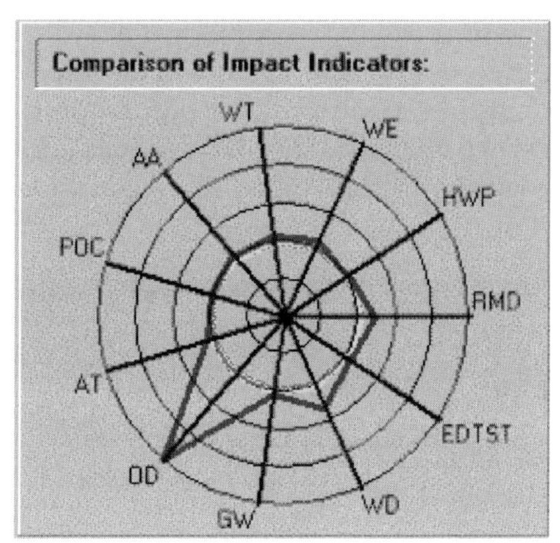

Figure 30 **Example of results from EIME™**

Figures 29 and 30 fully illustrate the philosophy of the tool: the product is quick to describe, the evaluation is easy to perform, and makes comparison with existing products, reference targets or customers' requirements straightforward.

To give a more complete view of the tool, one has to mention the way qualitative information is handled. The tool is able to store qualitative information, basically as an extensive set of rules. These rules can be either regulations or internal rules, translating the environmental strategy of the company. This information is given back to the designer in the proper context, through 'to do' and 'warning' dialogue box.

Figure 31 gives a example of a 'warning' dialogue box, which will pop up on the screen only if the designer uses a material containing a heavy metal.

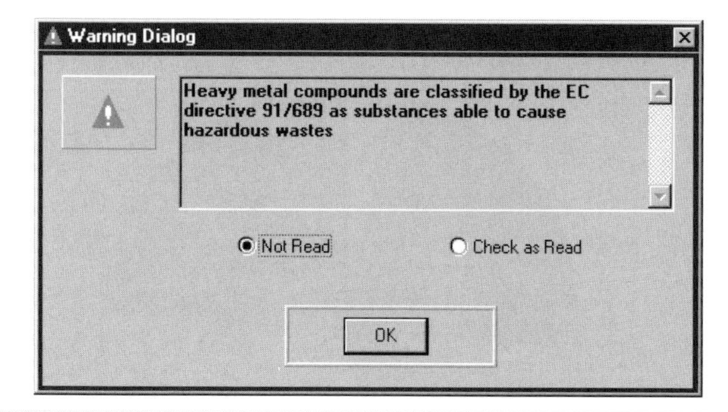

Figure 31 **Example of a 'warning' dialogue box in EIME™**

The expert has the possibility to customise many features of the software, and in particular:

- The modules used by the designers in the description of their design options
- The list of prohibited substances
- The indicators used by the non-experts for their evaluation process
- The warning/to do dialogue boxes, and the associated rules/context
- The list of compatible substances for recycling

This demonstrates that the tool offers a structure for a DfE approach, and that each company must specify it, in order to make it efficient, to adapt it to its own policy, information structure and project management scheme.

EIME™ database

The content and the quality of the database is of primary importance for the quality of the results, since designers are not going to complement missing data with original specific data.

The content of the current EIME™ database is described in Figure 32. It contains about 180 modules. These modules cover classic LCA data, end-of-life data and toxicology data. While electronic parts are well represented, other more generic modules such as materials and processes are also present, which could be used in other industries, in particular in the automotive industry.

In order to validate the quality and relevancy of the data stored in EIME™, an external validation procedure, involving one technical expert and one environmental expert per module, was defined and carried out on 50 modules in 1998.

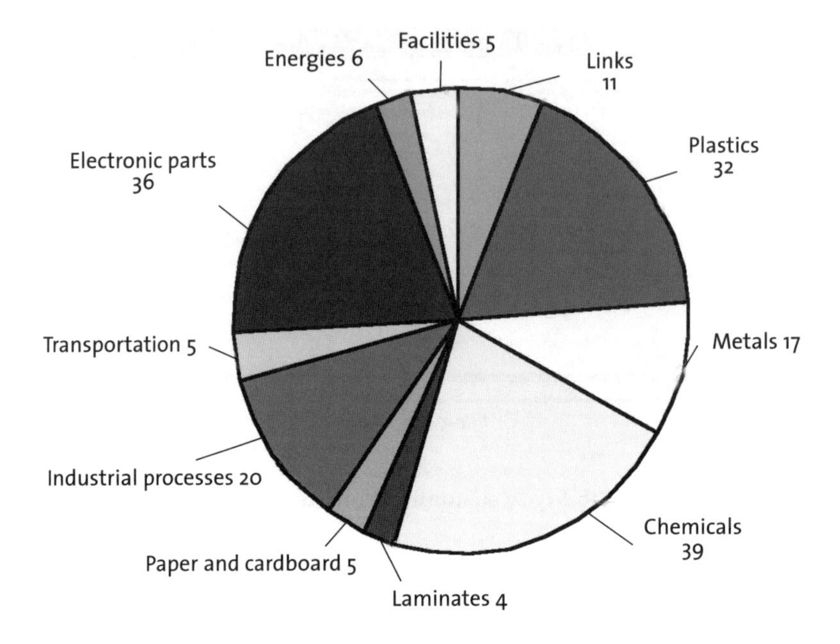

Figure 32 **EIME™ database content: number of modules per subject (December 1998)**

Room for improvement

DfE is a process, and its support tool, EIME™, is by definition evolutive. Three directions are being developed today:

- Improvement of end-of-life modelling: the idea is to feed EIME™ with results derived from other existing tools, specialised in end-of-life modelling (waste management, incineration, dismantling).[21]
- Development of procedures for the integration of DfE into a given company: implementation patterns are being explored and identified so that the DfE approach meets the constraints of the project management structure in different companies. This also implies building bridges between the DfE tool and existing databases, dealing with the definition of the product (dimensional definition), the environment (list of forbidden substances), the design process (existing guidelines). The objective here is to integrate the DfE approach into existing tools and methods, in order to ease its adoption by the company. Actual case studies are being developed, in co-operation with a research centre in management.

21 Ecobilan has, for example, a tool dedicated to the management of waste, Wisard™, and a tool dedicated to the modelling of the incineration process. Results from these tools could be integrated into the EIME™ database.

- Extension of the approach towards the suppliers ('full chain suppliers'): the objective is to explore how different companies in the same chain can exchange information and co-operate in the DfE approach. This is particularly important today, when larger and larger design responsibility is transferred from original equipment manufacturers (OEMs) to their suppliers. Some suppliers have already identified such a tool as a new vector for the marketing of their product. For example, 3M developed a label that is compatible for recycling with the most common plastics. A new module corresponding to this new label was created and implemented in the EIME™ database, to make it available to designers through the software tool.

Conclusion

The development of the EIME™ project with the electronic industry demonstrates the interest and the feasibility of a large multi-client DfE programme for complex products. It offers the opportunity, for other industries, to share a three-year experience, and to build on an existing structure. Whatever the method chosen to implement DfE in other industries, it is clear that the DfE process will be company-specific: the environmental rules, strategy and targets of each company are different, as well as the information organisational scheme and project management schedule.

There is interest in programmes gathering several companies (suppliers and clients) around a common DfE platform. The objective could be to design a highly modular DfE tool, which would allow the creation of a common language in the relatively new field, and the reduction of costs for each participant. Later on, the common platform can be adapted and tailored by each company to its specific needs and expectations.

ANALYTICAL TOOLS

Product development

Dr Todd Werpy and his colleague Ken Humphreys work at the Pacific Northwest National Laboratory, which is operated by Battelle. About 5,000 people work at the lab in Richland, WA, which specialises in basic science and technology. At different times in the past, the Pacific Northwest National Laboratory has been famous for its research and development leading to the compact disc, the photocopy machine, and the copper-clad coin, among other achievements.

Werpy and Humphreys work in the creation of intellectual property, which is then licensed or sold to others for commercial development. The pair are top-notch in their presentation and communication skills as well as in their science—skills that are important in a field where the important thing is convincing your audience of the inevitability of what you are saying. They travelled to Paris in 1999 to make the following presentation, which made use of slides to accompany their joint oral presentation.

INTEGRATING ENVIRONMENTAL CONSIDERATIONS INTO PRODUCTS AND PROCESSES

Todd Werpy and Ken Humphreys

We are interested in how to make strategic decisions when it comes to identifying economically attractive products of the future that will help achieve a sustainable world. Without an economic driver, the most environmentally benign of solutions will not be successful in the marketplace. On the other hand, early business success with sustainable products and processes will only broaden industry's desire for more of the same.

The future industrialised society will operate as depicted in Figure 33.

Currently, the petrochemical world is missing out when it comes to emulating nature. So the question we asked was: what are the alternatives for the petroleum-based C2–C6-derived products? What if we take, as raw material, corn, soy beans, wheat, livestock waste, food-processing waste and wood pulp, instead of petroleum? With all of these, you have starch, cellulose and glucose, and, with these, you can get to the basic platform chemicals (see Fig. 34).

The chemical industry in the US uses 900 million barrels of oil a year, and you can replace all of that by doubling the production of corn on existing acreage. Our process is the stage-gate approach, where we come up with the best-case economic scenario. If we find an idea doesn't work at any stage, we don't do any research and development on it (see Fig. 35).

Let us look at succinic acid. With succinic acid, you can make between 30 and 50 compounds, so here we selected 1,4 butanediol (BDO) and compared the petroleum-based and the bio-based systems (Figs. 36–38). We found the production costs equivalent, and the CO_2 emission, was identical for the two, petroleum-based and bio-based (see Table 5).

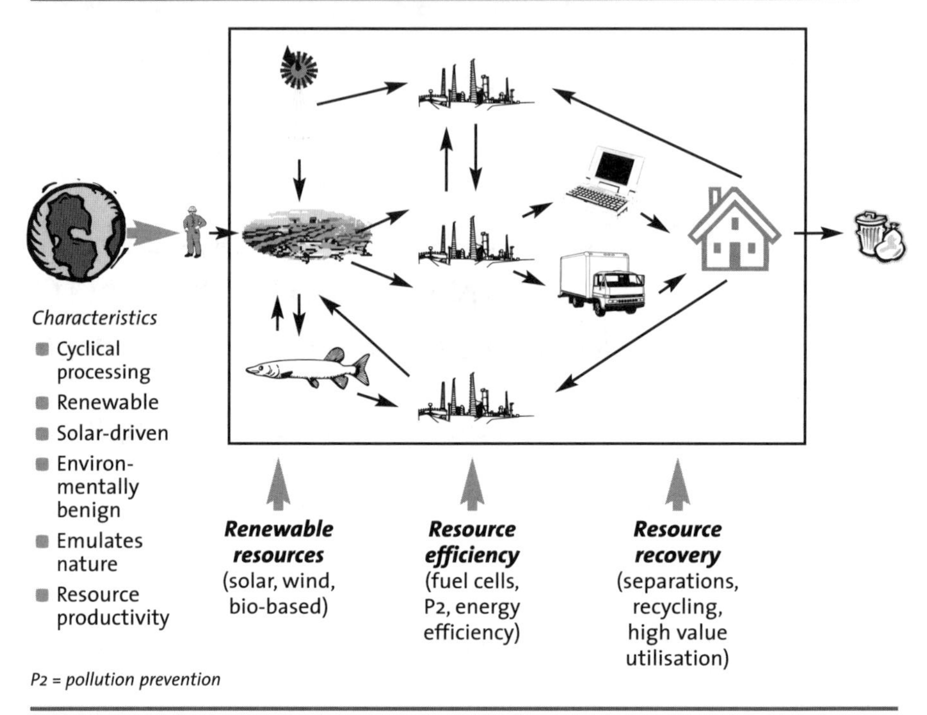

Figure 33 **Future industrialised society: cyclic, renewable system**

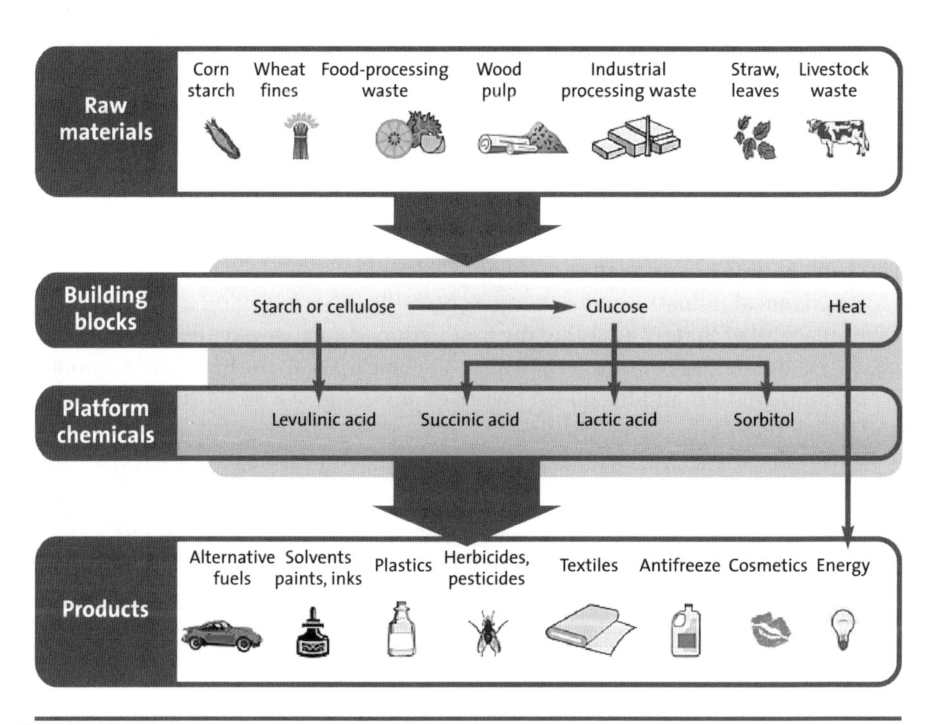

Figure 34 **A place to start: a bio-based future that complements our petroleum economy**

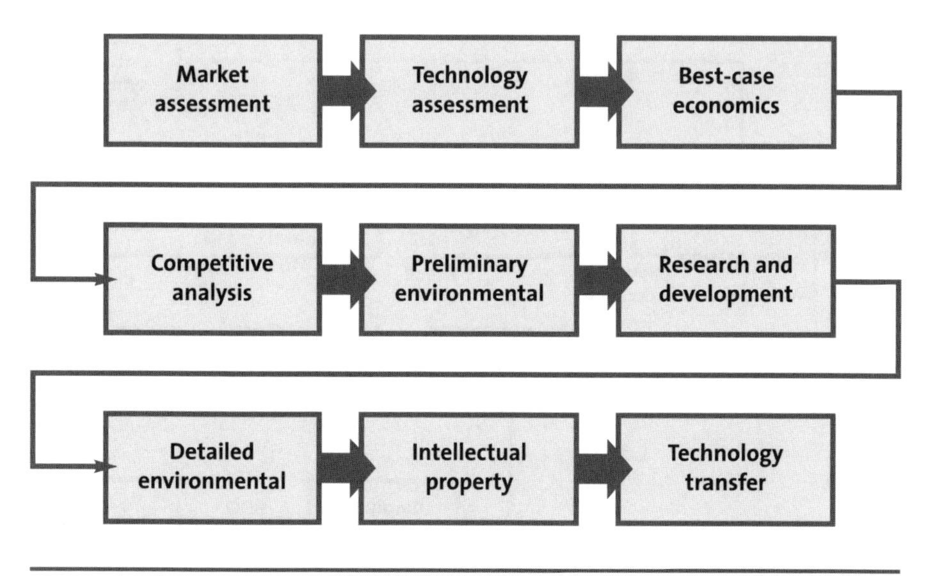

Figure 35 **Sustainable technology: a stage-gate approach to technology development**

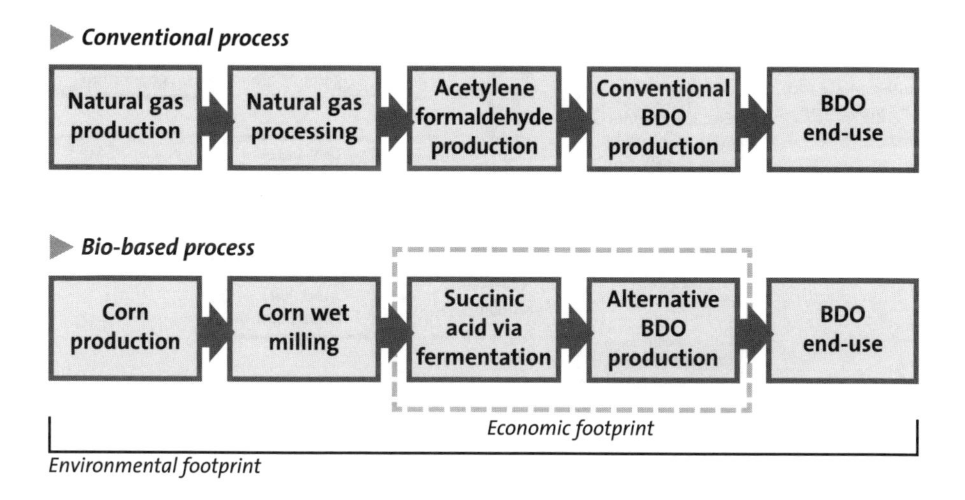

Figure 36 **1,4 butanediol (BDO) system: petroleum-based and bio-based systems**

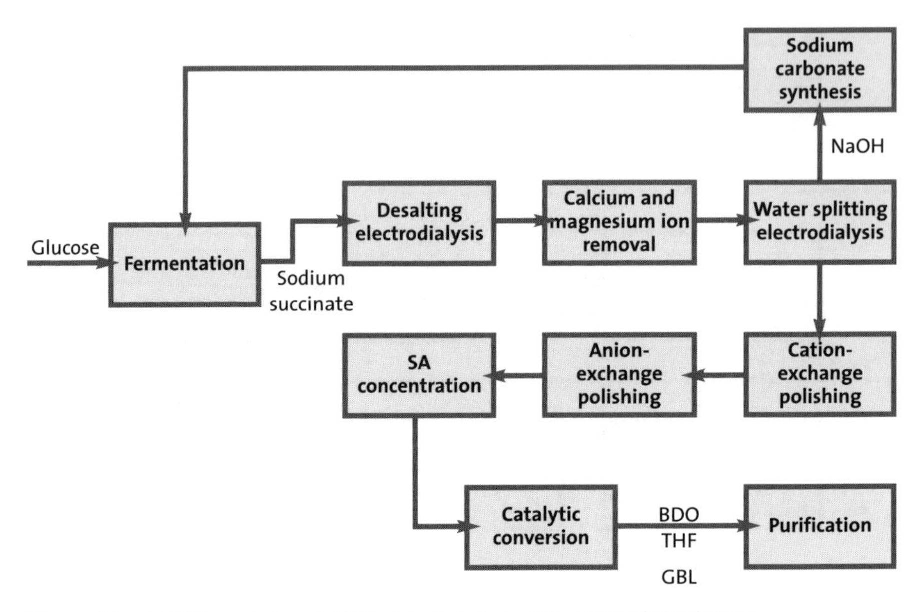

SA = succinic acid; BDO = 1,4-butanediol; THF = tetrahydrofuran; GBL = gamma butyrolactone; NaOH = sodium hydroxide

Figure 37 **BDO system: 1,4-butanediol manufacturing processes—baseline design**

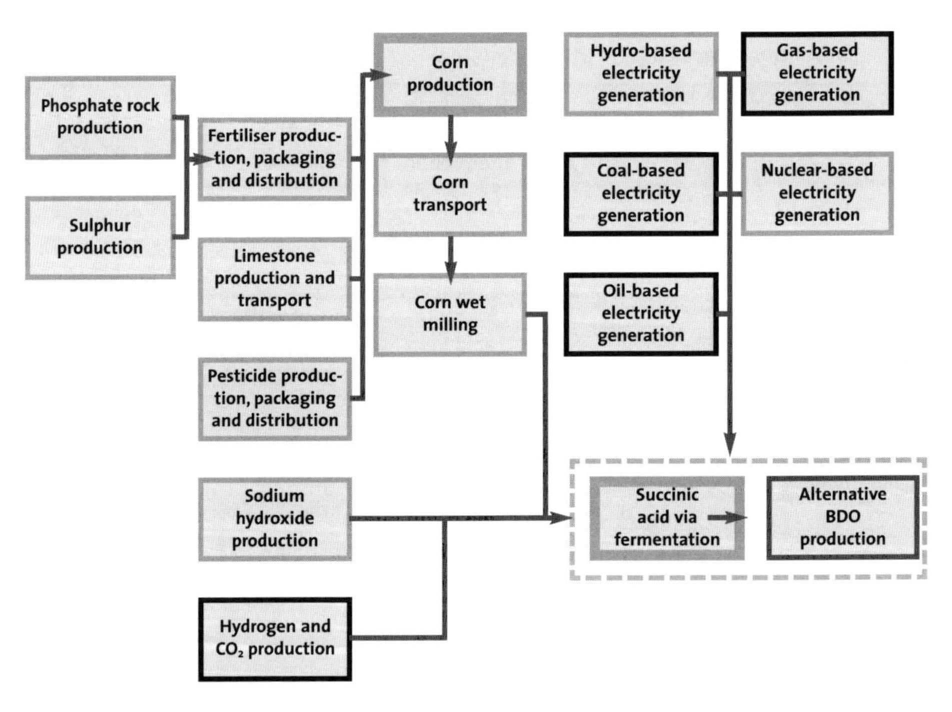

BDO = 1,4-butanediol; CO_2 = carbon dioxide

Figure 38 **Bio-based systems: environmental and economic impact—baseline design**

Impact category	*Ratio of bioprocess to petroleum baseline*
Production cost	1.0
Global warming (climate change)	1.0
Stratospheric ozone depletion	n/a
Acid rain	20.0
Eutrophication	1.2
Ground-level ozone ('smog')	1.5
Human health: chemical inhalation toxicology	3.4
Human health: PM inhalation effects	225.0
Human health: carcinogens	0.0006
Aquatic biota toxicity	6.5
Terrestrial animal toxicity	1.3
Resource depletion	0.5

PM = particulate matter

Table 5 **Bio-based systems: environmental and economic impact—baseline design**

Next, we asked, how can we make the corn-derived product better, environmentally and economically? We ran 200 sensitivity analyses for each economic model (Table 6 and Fig. 39).

Confirmatory visualisations
- Used to validate or test hypotheses about data
- Often emphasises quantitative correlation of variables

Exploratory visualisations
- Used to discern patterns in data
- May emphasise qualitative or quantitative aspects

Presentation visualisations
- Emphasis on persuading rather than explaining
- Often used for side-by-side comparison of options

Table 6 **Sustainable technology: using data visualisation to improve technology development**

You're always trying to eliminate useless information, trying to observe effects, validate hypotheses. Exploratory visualisation is, as much as anything, a way of sorting through piles of information. A presentation visualisation is a way of talking to everyone from researchers to the presidents of companies. A galaxy chart helps us to look across the chemical process design to identify, for example,

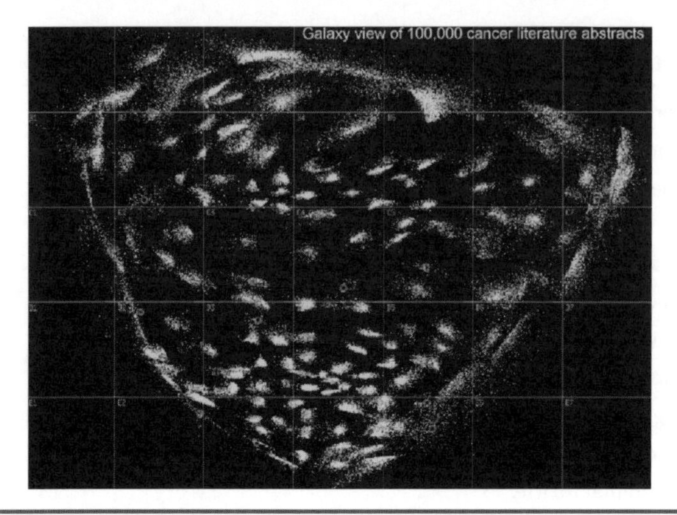

Figure 39 **Sustainable technology: exploratory visualisation—galaxy view**

fugitive releases from processes. A galaxy chart can represent hundreds of abstracts of scientific research that might show, for instance, a correlation between a chemical and cancer. These sensitivity analyses are a way of drawing on other fields of expertise to aid in the process of product design (Table 7).

▶ ***Web-based visualisation tools on company intranet***

- Document and communicate current economic and environmental 'status' of technologies under development
- Visualisations draw on a wide range of design tools
  - Question-based DfE screening approaches
    - Checklists
    - Software tools
  - Chemical process models
  - Life-cycle estimating and economy tools

Table 7 **Sustainable technology: presentation visualisation**

Sometimes we cast our net internally (Fig. 40). Pin-wheel diagrams communicate quickly in-house, and people can double-click anywhere for increasingly complex information.

So, to continue on with BDO, we came up with an improved design for 1,4-butanediol manufacturing (Fig. 41 and Table 8). We were able to find new enzymes for the fermentation process that excluded steps and cut 30% off the cost of manufacturing and substantially increased the environmental benefit. We got a much more environmentally beneficial product at 85% of the cost. The bio-

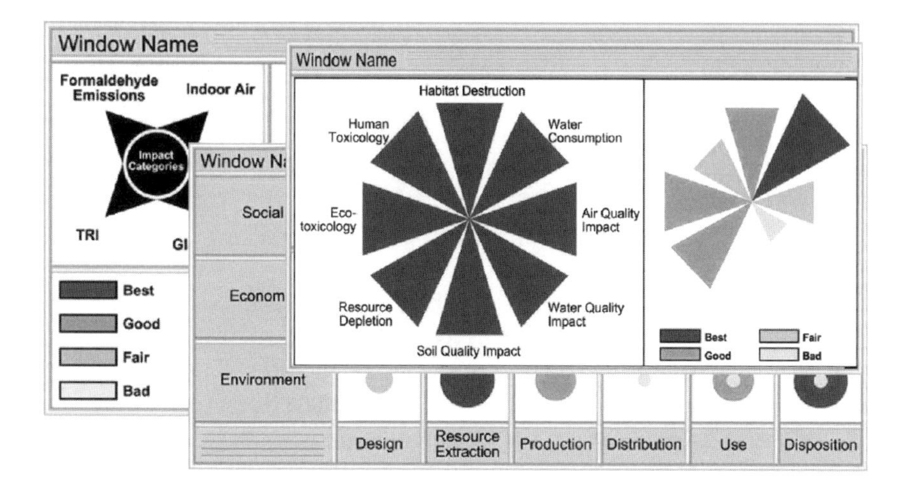

Figure 40 **Sustainable technology: presentation visualisation**

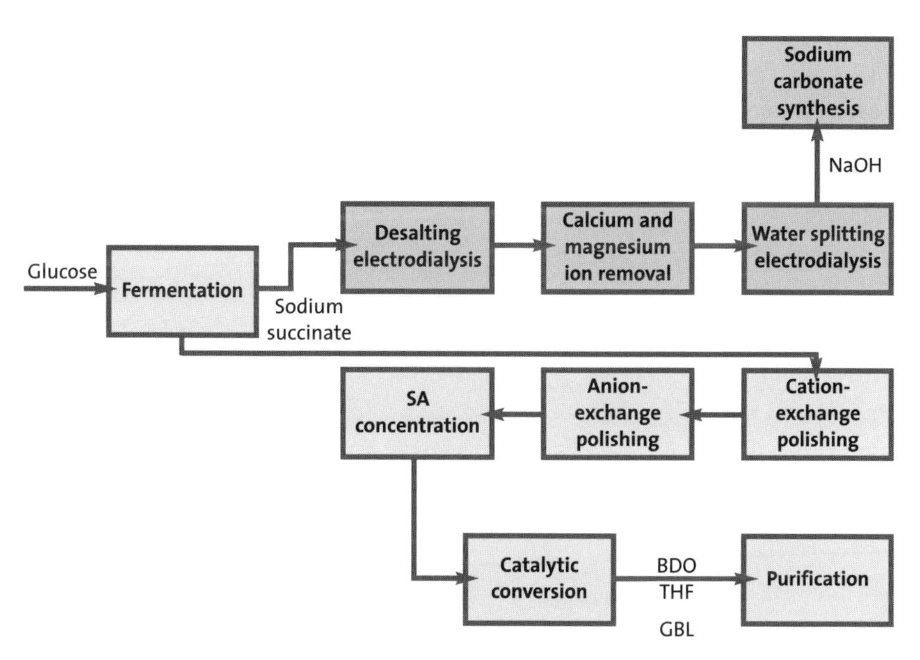

SA = succinic acid; BDO = 1,4-butanediol; THF = tetrahydrofuran; GBL = gamma butyrolactone; NaOH = sodium hydroxide

Figure 41 **BDO system: 1,4-butanediol manufacturing processes—improved design**

Impact category	*Ratio of bioprocess to petroleum*	
	Baseline	**2nd generation**
Production cost	1.0	0.85
Global warming (climate change)	1.0	0.38
Stratospheric ozone depletion	n/a	n/a
Acid rain	20.0	10.0
Eutrophication	1.2	1.2
Ground-level ozone ('smog')	1.5	0.75
Human health: chemical inhalation toxicology	3.4	1.0
Human health: PM inhalation effects	225.0	225.0
Human health: carcinogens	0.0006	0.0006
Aquatic biota toxicity	6.5	2.5
Terrestrial animal toxicity	1.3	0.86
Resource depletion	0.5	0.3

PM = particulate matter

Table 8 **Bio-based systems: environmental and economic impact—baseline design**

based product is much more attractive in terms of global warming effects, there is a reduction of toxic inhalations, making it one-five-thousandth as carcinogenic. It is obvious: the further back you can go in the research and development process, the greater the economic and environmental improvements.

FINANCIAL TOOLS

The financial world is an area of rapid change in the environmental landscape, with numerous new approaches to debt, equity, private equity and insurance. Maurice Strong, the main organising force behind Rio, said that the financial industry can have a greater impact on industrial environmental behaviour than all governments combined.

The remarkable success of the Grameen Bank in Bangladesh in the visionary **micro-credit** or **microfinance** movement is an excellent example of high impact. Working with the poverty–environment link, as well as with the removal of discriminatory treatment of women in the rural credit market, the Grameen Bank makes small loans, often to women, for projects or developments, often in agriculture, that are sustainable rather than destructive. As a mediator of economic, social and environmental change, micro-finance is a crucial factor in the sustainable development process. Institutions such as the World Bank, as well as Internet sites such as www.PlaNetFinance.org, are now implementing E.F. Schumacher's notion that 'small is beautiful' (Schumacher 1973). The World Bank is breaking new ground in channelling money directly to community groups, rather than to governments, with their micro-finance programme, 'Development Marketplace'.

For banks in general now, environmental performance may not be a reason to say 'yes' to projects, but bad environmental performance can now be a reason to say 'no'. Even conservative banks, such as the Dutch-owned ABN AMRO, contemplate involvement in the much-discussed carbon markets, which is, in a way, coming into environmental investment through the side door, but still coming in. British and German banks now have online systems addressing the environmental concerns of small and medium-sized businesses. There is talk that the traditional 'due diligence' analysis could in the future include environmental and social factors.

Environmentally friendly investment

The Bank Sarasin & Co. in Basel, Switzerland, funded a study called *Sustainable Investments: An Analysis of Returns in Relation to Environmental and Social Criteria*, in which 65 European securities were examined over a two-year period to measure their values against their social and environmental performance. While the study found no significant link between social responsibility and equity yields, it did identify a significant positive correlation in companies whose environmental performance played an important role in the public's perception (chemicals, pharmaceuticals, energy, construction, etc.). The study concluded that efficient use of resources and the avoidance of environmental risks help to reduce costs, and, while offering more environmentally compatible products, can boost earnings.

In Brazil, an investment bank called Banco Axial is at the forefront of 'biodiversity investing', a trend that began with socially conscious investors in Europe and the US. Axial's director, John Michael Forgach, is harnessing market forces to help protect Brazil's environmental assets, especially its rainforests, from degradation. Axial's Terra Capital fund can frequently leverage other investors, such as the Swiss government, the World Bank and other institutions and private investors, for such investments as a small sustainable timber operation in the rainforest, or a plantation producing carbon filters from coconut husks.

John L. Cusack is Chief Executive Officer of Innovest Strategic Value Advisors, Inc., with over 25 years' experience in the global environmental finance, management, products and services fields. He was head in North America and Europe of several environmental subsidiaries of Asea Brown Boveri (ABB), a multinational firm in which he served as adviser to venture capital funds as well as adviser to the corporate environmental affairs department. Since 1993, Cusack has specialised in management consultancy, in strategic environmental management, marketing and finance, working with industry, financial institutions and governments. He has an MS in environmental engineering and science, an MBA in finance and management, and is a registered professional engineer. The following is his take on the link between environmental and financial performance in businesses.

THE VALUE OF COMMUNICATING YOUR ENVIRONMENTAL POLICY TO WALL STREET

John Cusack

A 1999 PricewaterhouseCoopers study on 'The Information Reporting Gap in the US Capital Markets' found that 54% of investors believe that environmental performance is valuable, but approximately the same number feel that businesses do not do an adequate job of communicating about it. My company, Innovest Strategic Value Advisors, believes that superior 'eco-efficiency' is a powerful indicator for sustainable earnings quality and shareholder value creation. This value potential is not currently captured by traditional Wall Street analytics; thus an information arbitrage opportunity exists.

Our EcoValue '21™ analysis identifies the link between environmental and financial performance, a connection that will only increase as domestic and international regulations and discloser requirements tighten, and as new international performance standards and protocols (e.g. climate change protocol) come online. The EcoValue '21™ looks at 60 variables, including a company's history, operating risk, and improvement curve (Fig. 42).

The test: 162 stocks, drawn from the Standard & Poor's 500, were given EcoValue '21 ratings. The results over the three-year period studied are shown in Figure 43.

Sometimes hidden values are revealed by our analysis. Mobil Oil and Unocal carry virtually identical risk ratings from Wall Street, but the EcoValue '21 analysis reveals that they are far apart in both their eco-efficiency levels and their prospects for sustainable earnings quality (Fig. 44).

In high-impact sectors, the eco-efficiency premium is even greater than it is in more broadly diversified portfolios (Fig. 45 and Table 9; Fig. 46 and Table 10).

Critical issues for our analysis include how the company integrates its environmental strategy into its management strategy, how new services/products are

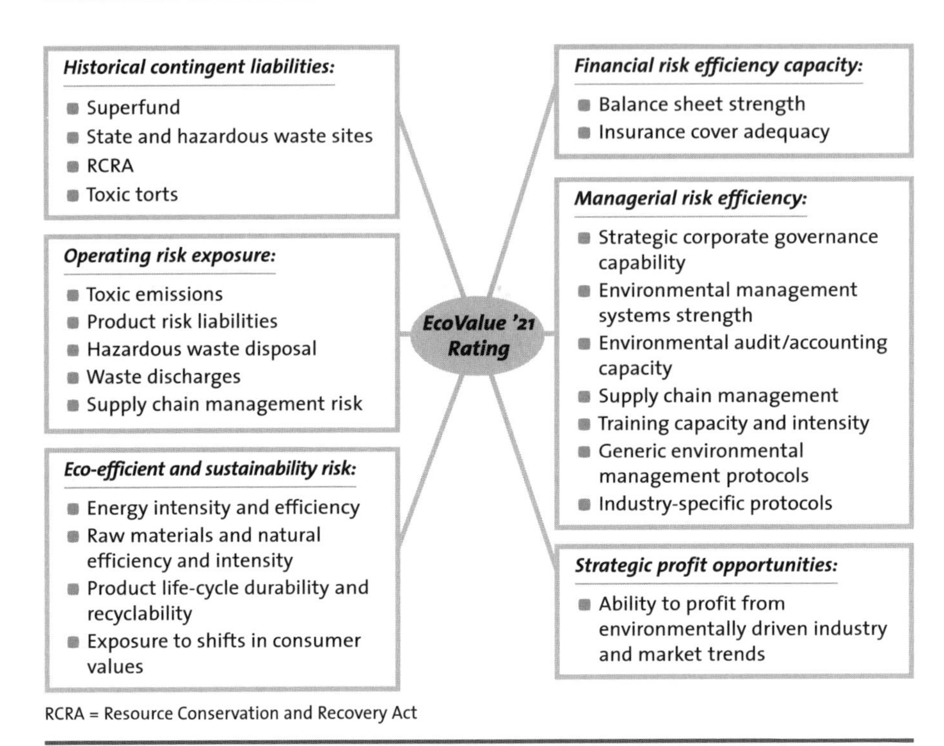

Figure 42 **EcoValue '21™: the investment analysis process**

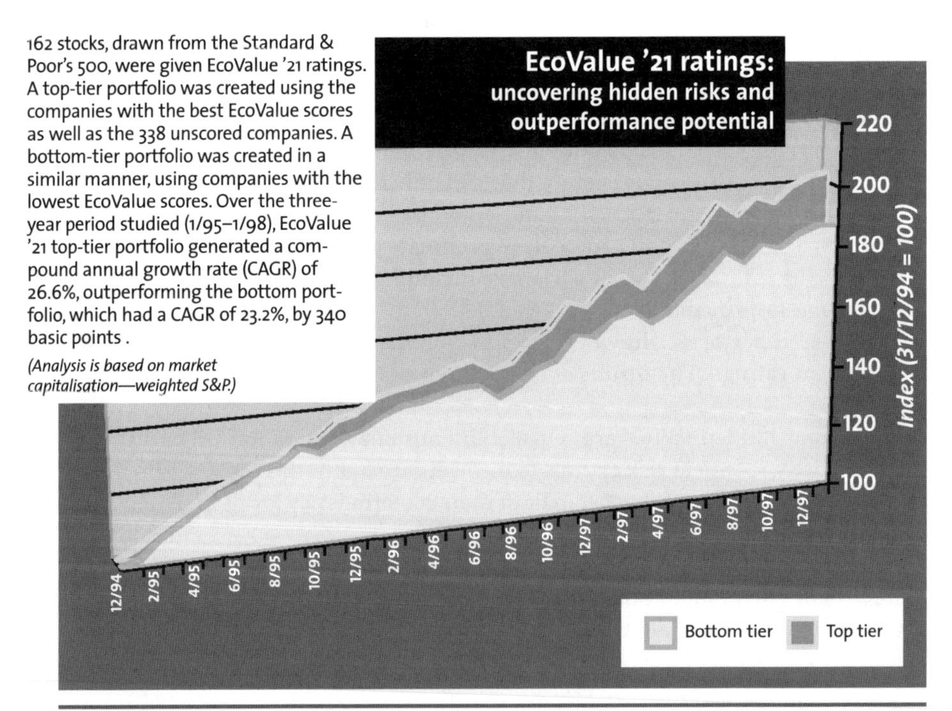

Figure 43 **EcoValue '21™ results: three-year backtest**

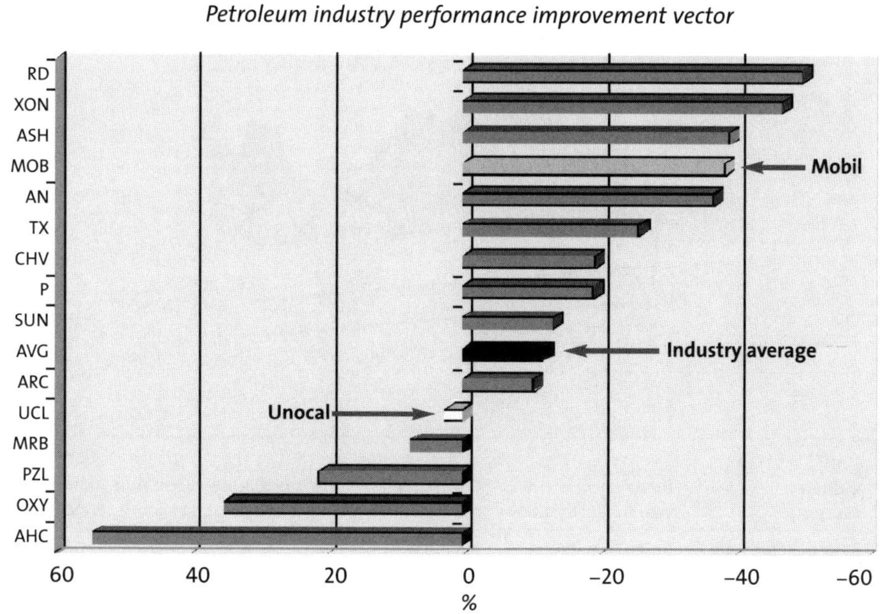

Petroleum industry performance improvement vector

RD = Royal Dutch; XON = Exxon; ASH = Amoco Shell; MOB = Mobil; AN = Antrim Energy; TX = Texaco; CHV = Chevron; P = Petrofina; SUN = Sunoco; AVG = Average; ARC = Arco; UCL = Unocal; MRB = Malaysian Rubber Board; PZL = Pennzoil; OXY = Occidental; AHC = Amerada Hess

Note: Mobil Oil and Unocal carry virtually identical risk ratings from Wall Street, but the EcoValue '21 algorithms reveal that they are poles apart in their eco-efficiency levels and their prospects for sustainable earnings quality.

Figure 44 **EcoValue '21™ results: hidden value revealed**

planned, the long-term health effects of products, and global warming impacts of the company's activities. Innovest's EcoValue '21™ environmental rating system has been used, starting in January 2000, as the investment strategy for an Eco-Enhanced Index Fund marketed by a subsidiary of Mellon Bank. This Eco-Enhanced Index Fund essentially tracks the S&P 500 Index within ±150 basis points, yet over-weights the firms highly rated by EcoValue '21™ and under-weights the firms that have lower EcoValue '21™ ratings, a strategy that produces financial returns consistently higher than the S&P Index itself.

EcoValue ’21
Top- and bottom-rated companies: 1998 total return

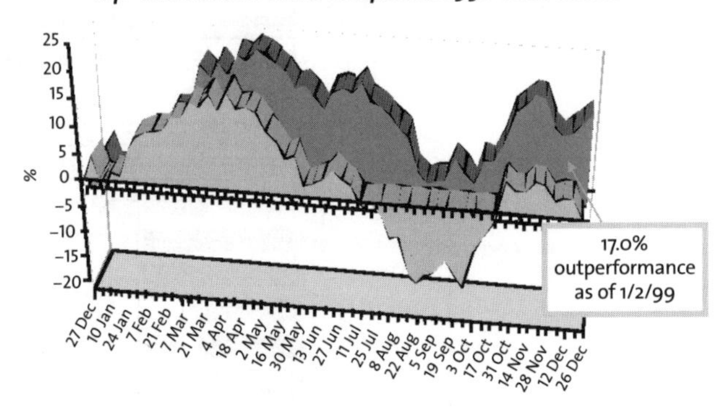

Industry category	Best/ worst	Symbol	Company name	EcoValue ’21 Score	EcoValue ’21 Score	S&P common stock ratio
Aerospace/defence	Best	BA	Boeing Co.	1,407	AAA	B+
	Worst	GD	General Dynamics Co.	912	CCC	B
Chemical: specialty	Best	ECL	EcoLab Inc.	1,585	AAA	B
	Worst	IFF	Intl Flavours & Fragrance	800	CCC	A+
Chemicals	Best	DOW	Dow Chemical	1,510	AAA	B
	Worst	FMA	FMC Corp.	1,015	CCC	B
Communication equipment	Best	NT	Northern Telecom Ltd	1,794	AAA	B
	Worst	HRS	Harris Corp.	1,073	CCC	B+
Electric companies	Best	POG	Pacific Gas & Electric	1,685	AAA	n/a
	Worst	FE	First Energy	645	CCC	B
Electric: semiconductors	Best	INTC	Intel Corp.	1,529	AAA	B+
	Worst	MII	MicronTechnology Inc.	1,033	CCC	B
Healthcare	Best	JNJ	Johnson & Johnson	1,546	AAA	A+
	Worst	MKG	Mallnckrodt Group Inc.	681	CCC	B
Iron and steel	Best	IAD	Inland Steel Industries	1,365	AAA	B–
	Worst	BS	Bethlehem Steel Corp.	1,015	CCC	B–
Paper, forest products and containers	Best	GP	Georgia-Pacific Corp.	1,616	AAA	B
	Worst	PCH	Potlatch	925	CCC	B+
Petroleum	Best	TX	Texaco Inc.	1,601	AAA	B
	Worst	PZL	Pennzoil	1,057	CCC	B+

Notes:

- In high-impact sectors, the ‘eco-efficiency premium’ is even greater than it is in more broadly diversified portfolios.
- The EcoValue ’21 analytics platform uncovers value potential overlooked by conventional investment analysis. Despite their virtually identical ratings from Wall Street, these pairs of companies from ten major industry sectors have radically different EcoValue ’21 ratings and investment performance.

Figure 45 and Table 9 **EcoValue ’21™ results: ten best and ten worst portfolios**

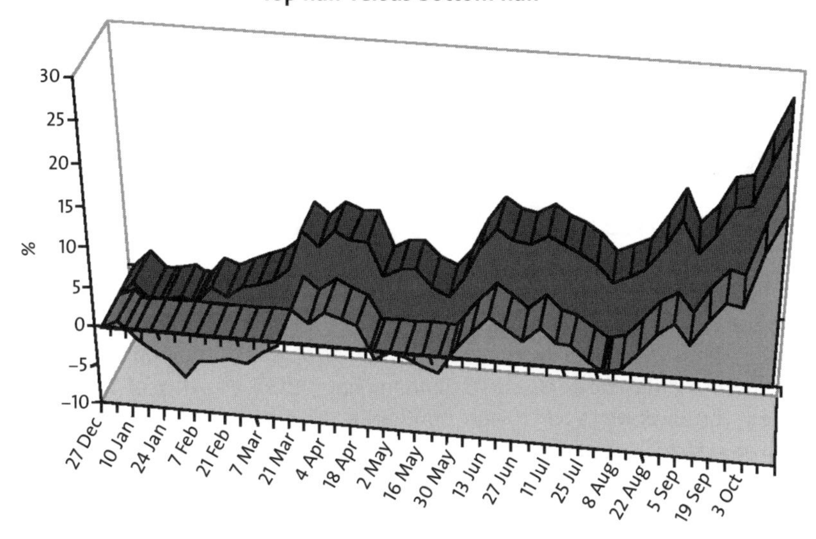

	27 Dec	1 Jan	17 Jan	24 Jan	7 Feb	14 Feb	28 Feb	7 Mar	21 Mar	28 Mar	4 Apr	18 Apr
Bottom half (%)	0.0	1.0	–2.4	–3.4	–3.4	3.2	–3.0	–1.6	4.4	2.9	4.5	2.9
Top half (%)	0.0	1.8	0.0	0.6	2.0	1.2	2.9	3.7	10.2	8.4	10.5	9.4

	25 Apr	9 May	16 May	30 May	6 Jun	20 Jun	27 Jun	11 Jul	18 Jul	25 Jul	8 Aug	15 Aug
Bottom half (%)	–0.9	–0.5	–1.3	1.1	3.5	4.4	2.8	2.9	2.7	1.0	0.7	2.6
Top half (%)	5.2	6.4	4.5	5.8	9.6	11.2	10.7	11.0	10.2	8.9	7.9	9.0

	29 Aug	5 Sep	19 Sep	26 Sep	10 Oct
Bottom half (%)	6.6	4.0	9.7	9.3	20.5
Top half (%)	15.2	11.4	17.1	17.5	26.9

Figure 46 *and Table* 10 **EcoValue '21™ sample results: sector portfolio, 1998**

FINANCIAL TOOLS
Environmentally friendly investment

Tessa Tennant co-founded the UK's first investment fund for sustainable development in 1988. She has worked in the business ever since and stepped down as Director of Henderson Global Investors to pursue the ecological transformation of the capital markets from a different perspective. Tennant has served as Chair of the UK Social Investment Forum as well as on the environmental advisory panel for HRH The Prince of Wales. In 1997, she initiated the Asia Pacific Research project to foster social investment in the region, and the resulting Global Care Asia Pacific Fund was created in 1998.

Tennant's voice is that of an investment broker—very goal-oriented—but her goals differ from the norm. Tennant has not waited for the world's agencies to agree on or define sustainability indicators in order for her to determine where investments should be made. She has defined an investment methodology based on her own sound instincts, which has been enormously successful. The following remarks were made in Paris in 1999 when she was with NPI Global Care.

MAKING AN INVESTMENT IN YOUR FUTURE

Tessa Tennant

> Pension funds must consider how their funds are invested. The government believes that, subject to the overriding requirements of trust law in respect of the interests of beneficiaries, trustees should be able to consider moral, social and environmental issues in relation to their investments (Department of Social Security 1998).

NPI Global Care Investments established its first eco fund in 1988. We have had such success with our efforts that we are starting one in South-East Asia. Global Care is part of the AMP Group, a team managing £70 billion, with eight million policy holders. It is probably obvious to anyone that, with this kind of money, you can have a substantial effect on global business policy: you can invest in companies that contribute to, benefit from and adapt to the global trend towards sustainable development (see Fig. 47).

Our research methodology and rationale is as follows:

1. We identify industries of the future by looking for or at
   - Value-added knowledge
   - Sectors benefiting sustainable development
   - Regulatory, consumer and geographic trends
   - Insights acquired through contacts with stakeholder groups
   - Advisory committee expertise

2. We are particularly interested in clean energy (Fig. 48).

3. We look for clean engine technology (Fig. 49).

4. We look for businesses or institutions that are 'best in class', that have a vision for sustainable development, that have environmental management systems in place, that practise product stewardship and full disclosure, and that concern themselves with the environmental performance of their supply chain.

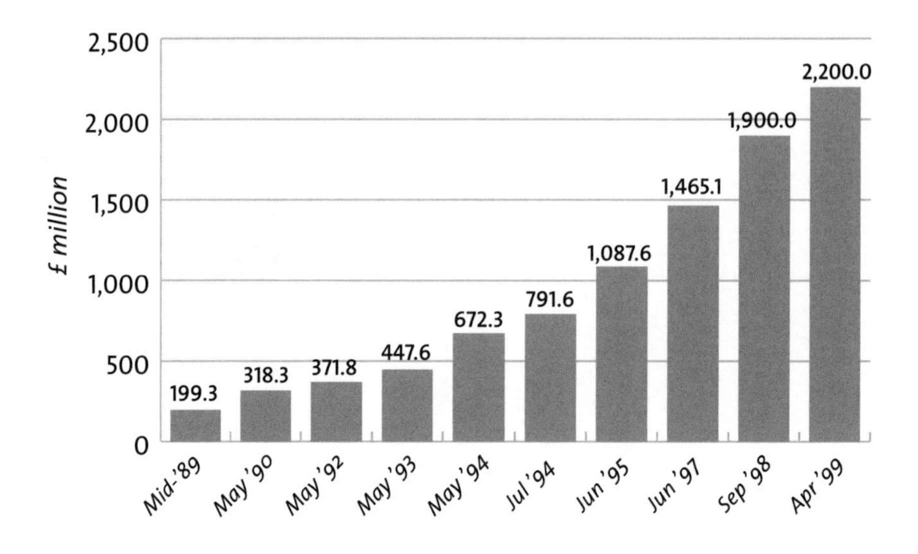

Figure 47 **Social investment: set for take-off?**

Source: Ethical Investment Research Service (EIRIS), April 1999

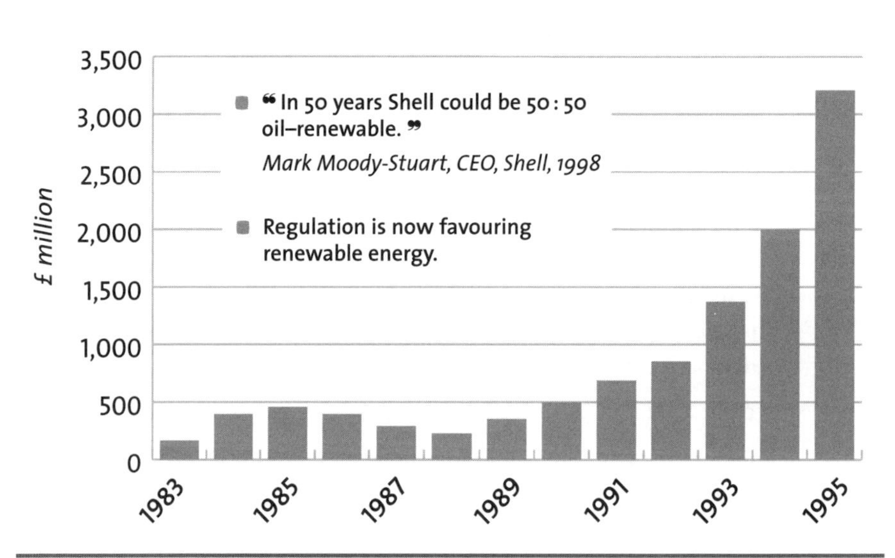

Figure 48 **Global installed wind capacity**

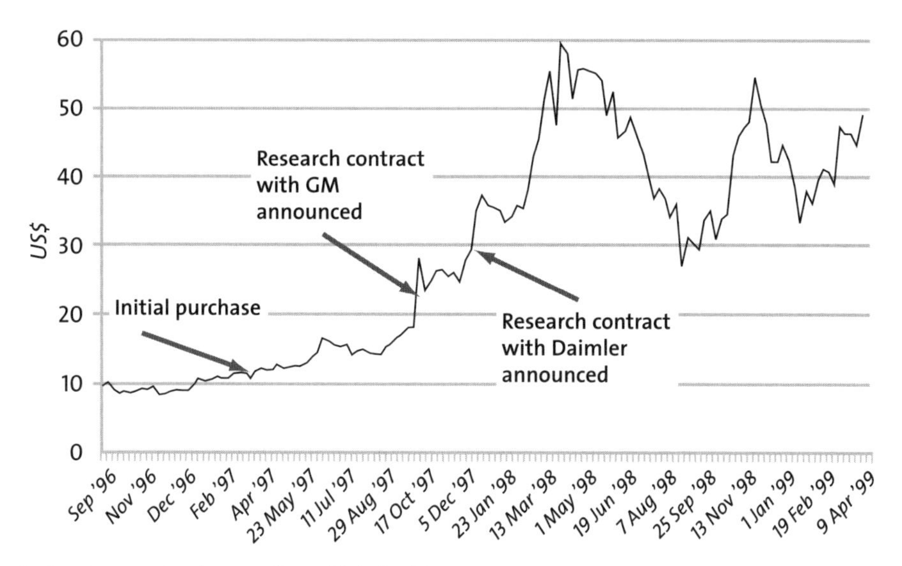

Figure 49 **Share prices, Ballard Power Systems: fuel cell technology to reduce vehicle emissions**

Source: Datastream ICV

There is one major problem in our assessments, but we imagine that it will be resolved shortly: companies have no agreed-upon standard for calculating their CO_2 emissions. The question is being examined worldwide, however, so perhaps by the time you are reading this the problem will have been solved.

Editor's note. In 2001, the GHG Protocol Initiative proposed a standard for measuring and reporting business greenhouse gases.

In terms of best in class, in the category of European Banks, we looked at environmental practices, their management of third-world debt, their human rights record, community involvement, business practices and so on. The best performers are five Spanish banks, three German banks, three UK banks, one Danish bank and one Irish bank. There were no French or Italian banks that measured up to our best-in-class standards.

In our research and investment-management process, we cast a wide net. We rely on contact with non-governmental organisations, contact with overseas research agencies, specialist reports, online media, public registers, analyst reports, analyst meetings, macroeconomic reports and internal investment research.

The all-important bottom line is that the NPI Global Care Growth Fund is up 192.3% since its launch in 1992, while the average international growth sector is up 169.5%.

FINANCIAL TOOLS
Public–private leveraging

The World Bank, according to Ian Johnson, Vice President for Environmentally and Socially Sustainable Development, must help create a sustainable future. Before the concept of sustainable development, World Bank progress indicators reflected society's past achievements, whereas sustainable development indicators must anticipate what society desires for its future (the quantities of greenhouse gases emitted by national economies based on fossil fuels is a sustainable development indicator, for instance). In a world whose population will double in the next five years, the World Bank cannot repeat its past mistakes, where money awarded to a country often ended up in the wrong hands or did not benefit those in need. Johnson says that, now, when the Bank makes misjudgements, they hear about it on the Internet. Now, he says, they have to get money to people rather than to governments.

In 2000, the World Bank started a venture designed to create a market for buying and selling rights to release pollutants that have been blamed for global warming. According to the Kyoto Protocol rules, companies in developed countries will be able to meet obligations to reduce carbon emissions by buying offsetting reductions in less energy-efficient countries, where reductions can be achieved more cheaply. The Bank feels the venture is consistent with its goal of alleviating poverty, because the world's poor often live in places that are most vulnerable to damage from global warming—low-lying coastal regions and areas where agriculture is marginal and medical treatment unaffordable. The World Bank venture is funding projects now, in the hope that the know-how acquired will be available later, when and if Kyoto comes fully online. The Bank believes they can reduce carbon emissions in poor countries for $20 a ton or less, as compared with costs of more than $50 a ton in already energy-efficient developed nations. One of the first projects, for instance, will reduce the methane gas produced by 27 open dumps in Latvia by creating a regional waste-treatment facility, increasing recycling and capturing methane gas to produce electricity.

Louis Boorstin is Manager of the Environmental Projects Unit in the Environmental Division of the International Finance Corporation, a member of the World Bank Group. Previously, he managed World Bank activities in Vietnam, where he created a leasing sector crucial for development. Boorstin apparently likes the challenge of developing new approaches that facilitate large change. He made the following remarks to explain his unit in the International Finance Corporation in Paris, in 1999.

REMARKS

Louis Boorstin

The International Finance Corporation (IFC) is part of the World Bank effort to promote sustainable development through encouraging and aiding private-sector investment in developing countries. The objectives, quite simply, are to reduce poverty and improve people's lives. The IFC is only four years old and its mere existence is indicative of the change occurring in the financial sector when it comes to sustainable development. The IFC can serve as a catalyst to stimulate and mobilise private investment where it might not otherwise happen. As an honest broker, the IFC can be a reassuring presence for joint-venture partners and host governments in the developing world.

So far in our brief existence, the IFC has invested US$3.5 billion, matched with $2.5 billion from private banks and $10 billion from project sponsors. The IFC is the risk-taker and guarantor. Projects must lead to further development within the country, and we undertake only commercially viable projects or, as is often the case with environmental projects, projects that are close to being commercially viable. This exceptional condition is because environmental projects typically have longer lead times, with newer technologies and less experienced sponsors.

The IFC can offer project financing, both debt and equity, mobilisation of capital from other sources, financial market development, and all sorts of advisory services.

IFC projects have been in clean water (Binh An, Vietnam Bulk Drinking Water), waste management, energy efficiency, renewable energies, including biomass, wind, solar (there is currently a joint IFC–Global Environment Facility project called the Photovoltaic Market Transformation Initiative to accelerate use of solar power in India, Kenya and Morocco), geothermal, and small-scale hydro, eco-tourism, sustainable forestry and agribusiness (a hearts-of-palm business in the rainforest where the forest is not damaged because the canopy is necessary). The Global Environment Facility (GEF) has project funds separate from the IFC, and its projects must be in climate change, biodiversity or international waters issues.

The IFC is looking for good-quality, private-sector projects related to the Kyoto Protocol and the evolving market for greenhouse gas emissions. Following is a description of a project that the IFC could facilitate. If someone chose to proceed with a project of this description and wanted to obtain carbon offset financing, the IFC would try to assemble the package. This is their tender from 1999:

Urban solid waste management

Objective

The objective of the project is to reduce future greenhouse gas (GHG) emissions from waste disposal activities in an African city with a population of about 1 million.

Indicative project description

Over the next 20 years, the population of the city may grow by more than 50%. The city wishes to privatise its waste management function. Several alternatives have been evaluated but none are commercially appealing. One alternative, to develop refuse-derived fuel (RDF), could become appealing to an investor if emissions reductions are valued and sold to an interested third party. Under the proposed RDF option, a local paper company would modify its boilers to combust RDF. This would avoid the creation of methane emissions from landfilling waste and would also displace emissions from burning coal in the boilers. Depending on the size and lifetime of the methane resource from the existing landfill, IFC may investigate potential uses for the methane (e.g. for waste management vehicle fleet).

Additionality

The baseline case is landfilling waste. The base case is to continue to use the current landfill and, once the current landfill is full, to open another further out of the city. The RDF option could be considered by the local paper company if it can be proven to be commercially viable. Without valuing the avoided methane and CO_2 emissions, the RDF option will not be pursued.

Status of baseline financing

The total cost of the baseline project is US$50 million. At this time, financing has not been secured because the project is not commercially viable. The city wishes to privatise the waste management function, and IFC is providing background material to support the consideration of GHG emissions reduction financing as part of the privatisation process.

Indicative matching contribution

The maximum allowed. Contributions from other sources may be sought.

Indicative emissions reduction
Ten million tonnes of CO_2 equivalent over 15 years. Emissions will be reduced over at least 20 years.

Indicative project processing timetable
The waste management function has not been privatised as yet, although discussions are under way. The attractiveness of any privatisation would be improved if the avoided methane and CO_2 emissions can be factored into the value of the future business.

Potential risks
The project that will eventually be undertaken by the proposed privatised entity is unknown at this time. A feasibility assessment indicates that the RDF option is the most viable from a commercial standpoint. However, even if the municipality proceeds with the privatisation, it is unclear which option the newly privatised entity will pursue (e.g. the landfill base case is cheaper). Nevertheless, the valuation and potential sale of emissions reductions could greatly influence the choice of project such that the project that reduces a large amount of emissions (e.g. RDF) will become more appealing.

FINANCIAL TOOLS

Environmental reporting

Business and industry in the 21st century must not only be competitive but accountable, according to the United Nations Environment Programme. Yet, while many companies now publish environmental reports, many of these same companies appear to lack a profound understanding of what sustainable development is or could be and, by implication, what it might really signify to have a triple bottom line, including economic, environmental and social value added.

The increasing attention focused on environmental performance of corporations has led to the new and controversial subject of environmental reporting. Stakeholders who might have an interest in the environmental activities of companies include: authorities, customers, employees, academia, non-governmental organisations, banks, national and international investors, neighbours, suppliers, contractors, municipalities, states, nations, competitors, future employees, media, insurance companies, financial analysts and political decision-makers.

Inevitably, the information in one company's environmental report is compared with information in another's. Thus, the problematical question of standardisation becomes important.

Novo Nordisk in Denmark is an example of a company that has helped develop the high standard for environmental reporting, and the company has won many prizes for its efforts. Some might say it is more well known for its environmental reporting than it is for its product line (among other things, it manufactures insulin). Its Director of Environmental Affairs, Lars Peter Brunse, calls environmental reporting the ice-breaker. Once you are into it, you have to talk about the not-so-good stories, or else.

Environmental reporting: a brief history

Lorraine Ruffing has a PhD in trade and development economics from Columbia University in New York and has worked in the fields of enterprise development, energy–economy interactions, and international accounting standards in a wide-ranging series of posts—at the University of Chile, the Indian Policy Review Commission in the US Senate, the UN Centre on Transnational Corporations in New York, and the UN office in Uzbekistan, among others.

Ruffing is presently Chief of the Enterprise Development Branch, Division on Investment, Technology and Enterprise Development at the United Nations Conference on Trade and Development (UNCTAD) in Geneva. Her branch specialises in the development of small and medium-sized enterprises in the global arena. International bureaucratic agencies burden their finest thinkers with impossible titles and labyrinthine divisions to manage, yet Lorraine Ruffing is a clear, concise communicator on the subject of environmental accounting. UNCTAD currently undertakes workshops on the subject in 25 countries. The following remarks are taken from a brief history of environmental accounting and reporting she presented in 1999 in Paris.

REMARKS

Lorraine Ruffing

Environmental reporting is a new, controversial subject with a very short history. The lion's share of business managers and CEOs still see no use for environmental reporting, which is of course a precursor to full corporate social responsibility accounting. Many professional accountants believe they have no role in anything outside financial accounting. Critics of this resistance believe that transparency is an issue, that most companies do not want their full stories known. For the United Nations Conference on Trade and Development (UNCTAD), which deals largely with developing and undeveloped countries, it has been difficult to establish to client nations that there is no trade-off in environmental protection.

The history started this way. After the Brundtland Commission and after the industrial accident at Bhopal, an international intergovernmental group took up the question of environmental accounting. There was much resistance. Why start a whole new game? And it remains the position of the accounting industry that environmental costs cannot be identified and separated out with any accuracy. It was the German chemistry industry that first separated out its environmental costs successfully, thus creating a model for what should and should not be disclosed in financial statements dealing with environmental costs.

The first surveys of the early environmental reports that emerged revealed that the information given by corporations about their environmental activities and costs was descriptive, partial and difficult to compare with information from any other corporation. In other words, the reports were anecdotal narratives. Further, the information given was neither in environmental terms nor financial terms, and thus made no sense to anyone.

In 1995, the international auditing firm KPMG undertook a survey of 800 companies: 19% disclosed environmental costs; 1% integrated this information into the body of their annual financial report; and 18% put the information in the small-print notes at the end (see KPMG's 1999 survey findings summarised in the editor's note opposite). In 1997, another survey made by the Associated Certified Chartered Accountants in the UK determined that environmental disclosures varied widely with no consistency in terms or methods.

The United Nations Environment Programme (UNEP)/SustainAbility project 'Engaging Stakeholders' looked at non-reporting of environmental information by businesses around the world, and the reasons given for non-reporting were: (1) doubts about the advantages of reporting; (2) an already-existing good environmental reputation which was at risk by reporting or presumably made reporting unnecessary; (3) the additional expense involved in reporting; (4) an unwillingness to potentially damage reputation; (5) doubts that reporting would increase sales; and (6) policy reasons. The project then came out with a list of 50 issues that they felt ought to be reported on (SustainAbility/UNEP 1998).

It was felt by many, including UNCTAD, that this wasn't helpful, because it was too ambitious and demanding, that they ought to start out more modestly. It is true, however, that indicators are still vague, that guidelines are necessary. Nevertheless, companies such as Monsanto forge ahead with their own model, and no one understands it or gives it credence.

It has been UNCTAD's project to synthesise all existing guidelines and to link environmental performance (which is not yet standardised) with economic performance (which is). Environmental performance indicators can look at environmental impact per $ of GDP, or per $ of sales, or for $ of profit and so on, and ratios can be established within a company, within an industry, within a country, etc. For UNCTAD, there must be a link between environmental costs and the financial bottom line, or there is no rational basis for future development. Establishing the parameters of this link is what UNCTAD is now working on.

Recent new definitions or areas of concern include the notion of 'constructive obligations', wherein environmental obligations derive from good business practices; and 'equitable obligations', wherein environmental obligations tend to be reported when they arise in developed countries where there is law and implementation of law, but not reported when they arise in non-developed countries where there is no legislation or no implementation. The World Bank has come out strongly in favour of making them both the same.

Now, just when needed, the Global Reporting Initiative has surfaced (see over) with continuing efforts to synthesise all existing guidelines and standards for environmental reporting. UNCTAD has joined with the GRI to this end.

Editor's note. In 1999, KPMG London surveyed 1,100 companies regarding environmental reporting. Of these, 269, or 24%, undertook environmental reports, with 35% of the top 250 companies surveyed filing environmental reports; 18% of the environmental reports published were externally verified, and 36% covered the subject of sustainable development; 53% of companies with environmental reports showed progress with their own targets.

FINANCIAL TOOLS
Environmental reporting

Global Reporting Initiative

The Global Reporting Initiative (GRI) was convened in 1997, not long after the *Exxon Valdez* incident, by the Boston-based group CERES (Coalition for Environmentally Responsible Economies), with the mission to design globally applicable guidelines for preparing business sustainability reports. There are over 150 organisations whose representatives participate in the long, complex and perhaps ultimately elusive process of standardising sustainability reporting. There are accountants, consultants, agencies, corporations, small businesses, banks, NGOs, lawyers, etc. struggling to elevate the standards to equal financial reporting standards, to bring about true 'triple-bottom-line' reporting, including environmental, social and financial information in one report.

The GRI vision is to support global progress towards sustainable development; the GRI Sustainability Reporting Guidelines will become the generally accepted, broadly adopted worldwide framework for preparing, communicating and requesting information about corporate performance.

The intent of the Sustainability Reporting Guidelines is to provide a framework that stresses the linkages between the environmental, social and economic aspects of enterprise performance. Sustainability at the enterprise level includes:

- **Environmental aspects**, including impacts through processes, products or services. These may include air, water, land, natural resources, flora, fauna and human health.
- **Social aspects**, including the treatment of minorities and women, involvement in shaping local, national and international public policy, and child labour and labour union issues.
- **Economic aspects** including, but extending beyond, financial performance, including activities related to shaping demand for products and services, employee compensation, community contributions and local procurement policies.

The process of the GRI now involves the attempt to agree on key sustainability indicators and to define them for their guidelines. A draft of the GRI guidelines was developed for public comment and testing through 1999. The guidelines are divided into nine parts, all of which are proposed as part of complete sustainability reporting. They are:

- CEO statement
- Key indicators
- Profile of reporting entity
- Policies, organisation and management systems
- Stakeholder relationships
- Management performance (with sub-parts pertaining to laws, conventions, and other mandatory standards; internal policies and standards and voluntary initiatives; external recognition and activities; suppliers; etc.)
- Operational performance (with sub-parts pertaining to health and safety; environmental performance; social and economic indicators; etc.)
- Product performance
- Sustainability overview

Twenty companies participated in the pilot-testing of the guidelines, including British Airways, Electrolux, Ford, General Motors, ITT FLYGT, Novo Nordisk, Procter & Gamble, Sasol, Shell, Sunoco, etc. As Magnus Enell, Corporate Environmental Director for ITT FLYGT in Sweden, said: 'It is big risk for us, for all of them. But if we start, others will follow.'

Countless organisations around the world already have or are involved in developing social and/or environmental accountability standards which the Global Reporting Initiative seeks to unify. An example of such an organisation is AccountAbility, the Institute of Social and Ethical Accountability, which has a system called AccountAbility 1000 (AA 1000).[22]

Another accountability standard is the Copenhagen Charter, a management guide to stakeholder reporting, developed by Ernst & Young, KPMG, PricewaterhouseCoopers and the House of Mandag Morgen in 1999. The Copenhagen Charter was launched at 'Building Stakeholder Relations: The Third International Conference on Social and Ethical Accounting, Auditing and Reporting', which took place in Copenhagen, Denmark, in 1999.

Another standard for corporate social accountability is SA 8000, developed by the non-profit Council on Economic Priorities Accreditation Agency with the assistance of experts from business and industry, labour and human rights organisations, and leading certification and audit firms from around the world. The SA 8000 looks at many aspects of corporate responsibility, including environmental, as fundamental business issues.

22 www.accountability.org.uk

TECHNOLOGY TOOLS

Technological advances can in themselves be tools to achieve superior environmental performance. Sometimes they come as a result of inevitable progress in technological innovation, sometimes by necessity, either governmentally encouraged or imposed, or, in the case of Sasol's environmentally friendly diesel fuel, geographically imposed.

An example of governmentally encouraged technological innovation (cleaner gasolines and more efficient engines) should result from the latest United States government's regulations regarding emissions standards for cars. The rules call for a 90% reduction in the amount of sulphur in gasoline by 2004, and emissions standards will apply to all vehicles, including pickups, mini-vans, vans and sport-utility vehicles which previously escaped tough emission standards. Implementing the new standards will be the equivalent of taking 54 million cars off the road.

In Japan, in order to meet their 2010 target for reduced CO_2 emissions, the government is expecting to obtain a 25% reduction through improved fuel efficiency or reduced fuel consumption in automobiles. The Toyota Motor Corporation is thus intent on meeting this requirement through continuously developing and honing its technological tools, as it is fully aware that its products contribute to global warming. Toyota has always been concerned with reducing waste as a means of increasing profits, and it has won many international awards for its efforts to eliminate wasted energy. In 1998, Toyota made the decision to require that all of the electricity used in its California operations would come from renewable resources, a choice made possible by California's deregulation of its electric utility industry.

As the following technical paper shows, Toyota is developing engines that use different fuels (natural gas, electricity, fuel cells), or that use fuel differently. An outsider's assessment of Toyota's proposals as described below would indicate that life-cycle assessments are needed to determine if environmental impacts will not simply be shifted from one place to another: in the case of electric vehicles, for example, from the place where the vehicle is used to the place where the electricity is created—that is, nuclear power or other electric power facilities.

It is interesting to note that Japan plans to meet its 2010 target for reduced CO_2 emissions through a 9% voluntary reduction in vehicle use, an ambitious and life-changing goal which, at the time of writing, no other country has put forth.

The following paper was presented in Paris in 1999, by Toyota's Satoshi Matsuura, who graduated from Kobe University in economics. After receiving his degree, Sotoshi Matsuura joined the Toyota Motor Corporation where he has been for 30 years. He was Toyota's General Manager of Environmental Affairs at the time of this presentation

TRENDS IN ENVIRONMENTAL ISSUES AND THE TOYOTA ACTION PLAN

Satoshi Matsuura

Automobiles are inexorably linked with the environment at every stage—manufacturing, use and end-of-life (recycling and waste). The impact of automobiles on the environment has been far-reaching—beginning with urban air pollution and acid rain and other problems that transcend local borders. And, now, emissions of carbon dioxide (CO_2) underscore the global nature of environmental problems. Measuring these emissions in terms of tons of carbon, we see that the equivalent of 3 billion tons is added to the atmosphere each year from combustion of fossil fuels and depletion of forests.

Japan's annual CO_2 emissions amount to some 300 million tons, or about 10% of the annual global total. Of that amount, Japan's automobiles account for 60 million tons, or about 2% of global total. We think these figures are significant. Therefore, we are doing everything we can to reduce them (and still make automobiles) because (1) we have a responsibility to make direct contributions to improving the global environment; (2) 'environment' now has a great deal to do with competitiveness in the world automobile industry; and (3) there is an increasing trend toward buying green.

The Toyota Earth Charter was adopted in 1992, calling on our management to address key environmental problems. The Earth Charter laid the foundation for our basic policies, action guidelines and plans. We have devised seven action guidelines and the Toyota Action Plan. We in the Environment Department promote external communications on environmental matters with such items as our Environmental Report.

The purpose of the Toyota Environmental Report is threefold. First, it is part of our efforts to provide timely and complete information on environmental issues. Second, it helps make Toyota employees more conscious of the environment, which will promote even better ways of dealing with environment-related issues. Third, it helps promote unified concern for the environment among dealers, suppliers and overseas affiliates.

Included in the Environmental Report for 1998 is background information on the current CO_2 situation. In the area of transport, in which motor vehicles account for some 90% of CO_2 emissions, CO_2 emissions of 58 million tons in 1990 will grow to a target of only 68 million tons in 2010. To reach that target, the Japanese government plans to reduce CO_2 emissions—25% of the reduction will be through improvements in fuel efficiency (based on the top-runner method [see below]), 5% by introduction and sales of clean energy vehicles, 44% by improvements in traffic and distribution systems, 9% by voluntarily reduced vehicle usage, and 2% by reducing unnecessary engine idling. Our research shows that 85% of the CO_2 emitted during a motor vehicle's life-cycle comes from its daily operation. That leads us to the conclusion that the best way to reduce CO_2 emissions is to reduce fuel consumption.

The top-runner method of measurement takes the most efficient current level of fuel use and sets an objective for improvement in that level, based on the assumption that technology will improve. The result is a very high standard, and it dictates the direction we auto-makers must take when considering ecodesign. Some of the projects that will help achieve our goals for reduced fuel consumption are direct-injection engines, continuously variable transmissions (CVTs), weight reduction and development of hybrid cars.

The second item in the Japanese government's CO_2 reduction programme is promotion of clean-energy vehicles. The government aims to have 200,000 electric vehicles (EVs) on the road by 2010, along with 1.8 million hybrid cars and 1 million natural gas-powered vehicles. Obstacles we must overcome include improving the economics of producing hybrid cars, development of high-performance, low-cost batteries, and developing leading-edge technologies such as fuel-cell EVs.

Ecodesign is a strong tool for reducing CO_2. Toyota is taking two approaches toward the development of power train technology. One focuses on improving conventional engines. The second is the challenge of developing new power train alternatives, using technology not yet available.

Keys to engine technology evolution include lean-burn engines, variable valve timing-intelligent (VVT-i) and direct-injection gasoline engines. New alternative power trains include EVs, hybrid systems and compressed natural gas. Further in the future, we expect to see practical fuel-cell EVs.

The Toyota D-4 engine, the direct-injection gasoline engine, was introduced to Japanese markets in 1996. Where conventional systems inject fuel into the intake ports, the D-4 system injects it directly into the cylinders, enabling much leaner combustion. In other words, the engine operates on much less fuel.

Conventional gasoline engines run on an air–fuel ratio of about 15:1. In the Toyota D-4, stable combustion is possible with mixtures as lean as 50:1. In that case, a huge volume of air must be sucked into the cylinders, so VVT-i is used to reduce pumping loss. Further, by minimising cooling loss of energy when the ultra-lean mixture combusts, we were able to greatly improve thermal efficiency, resulting in an engine with truly revolutionary fuel efficiency.

Compared to the same vehicle with a conventional engine, cars equipped with D-4 engines used 30% less fuel, based on an urban test mode, and showed 10% better starting and passing acceleration.

One hurdle remains in the development of Toyota D-4 engine—that of oxides of nitrogen (NO_x) emissions. Conventional engines combust a mixture with an air–fuel ratio of 14.6:1; this mixture permits simultaneous reduction of CO_2, hydrocarbon (HC), and NO_x emissions with three-way catalytic converters. However, conventional three-way catalytic converters do not function well with lean-burn engines such as the direct-injection Toyota D-4. But we developed a new NO_x reduction catalyst that solves this problem.

The NO_x-reducing catalyst stores NO_x until the engine load increases, such as during acceleration. Then the NO_x is broken down and released into the atmosphere as harmless nitrogen and oxygen. As of the end of February 1998, we have received patents on our NO_x storage alloy in ten countries around the world. We have either made agreements or are negotiating with a number of domestic and overseas auto-makers to supply this alloy to them. (However, we have a gentlemen's agreement not to disclose which auto-makers they are.)

As to alternative power sources, we introduced the Prius with the Toyota Hybrid System in December 1997. In this system, power from the engine goes to a power-split device, which divides the power into two paths: one to the wheels and one to the generator. Electricity from the generator can go directly to drive the motor or to charge the batteries. Depending on driving conditions, the system's computer decides whether to power the car with the motor only, with the motor and engine together, or with the engine alone. This minimises energy loss and greatly improves fuel efficiency. Furthermore, when the car decelerates, the motor acts as a generator, recovering energy that would be lost to friction or heat in ordinary cars and storing it in the batteries.

The Toyota Hybrid System boasts revolutionary fuel economy figures, not to mention amazingly low exhaust emissions. Compared to the Toyota Corolla, the Prius offers twice the fuel efficiency—28 km per litre in the Japanese city test mode. That means its CO_2 emissions are half those of Corolla. What's more, CO_2, HC and NO_x emissions from the Prius are just one-tenth the amount allowed by Japanese regulations.

As to electric vehicles, their major benefit is their zero-emissions operation. And if they are charged at night when electricity usage is low, they can help conserve a great deal of energy. Toyota markets an RAV4 EV, which has its batteries beneath the floor, ensuring ample passenger and luggage space. It uses newly developed nickel-metal hydride batteries and can travel over 200 km on a single charge. We feel it is a viable vehicle, largely because its new batteries store three times the energy of conventional lead-acid cells. These batteries are also used in the Prius's hybrid system. For now, EVs are most useful around local areas or in environmentally sensitive areas.

The fuel-cell EV is a new kind of power train that holds great promise for the future. The fuel-cell EV uses the chemical reaction between hydrogen and oxygen in the air to produce electricity, and the only emission is water. The key to the fuel-cell EV is a Toyota-developed hydrogen occlusion alloy that holds more than twice as much hydrogen as previous alloys. This allowed us to make the system compact enough to fit in a normal-sized vehicle.

Methanol is another potential fuel for fuel-cell EVs. It can be reformed on-board to provide hydrogen for the fuel cell.

There are still obstacles to overcome with fuel-cell EVs, including cost, vehicle weight and setting up a fuel supply infrastructure. Commercial application is still a long way off. No matter how often you point out that fuel-cell EVs reduce CO_2 emissions by about 70%, the cost is still high.

One last point—manufacturers should be aggressive in providing the information consumers need to buy green. Environmental programmes are a top priority at Toyota, even in the midst of Japan's ongoing recession. We have a plant in France where we will manufacture Yaris, one of our strategic cars for the coming years. We hope our project in France will serve to strengthen our countries' ties on environmental issues.

TECHNOLOGY TOOLS
The fuel cell

In the consuming world, the automobile is the symbol individuals identify with the most. People enjoy their cars enormously: the private space they provide, the fun or empowerment driving can deliver, the capacity to transport goods and other people, the door-to-door ease, and, most certainly, their car's aesthetic statement and the assimilation of this statement into their own image. Even if society moves increasingly toward public transport (which is not yet the case), the automobile will long remain the choice of many and must, thus, become a more environmentally acceptable choice.

Seven out of the ten large automobile manufacturers are now investing in the fuel cell as the environmentally sound choice for the future. Ballard Power Systems Inc., based in Vancouver, Canada, is the world leader in the development and manufacture of proton exchange membrane (PEM) fuel cells. Ballard is working in collaboration with the Ford Motor Company, DaimlerChrysler AG (these first two are part owners of Ballard), Alstom, Ebara and Toyota, to make the emission-free automobile a reality in the near future. By 2004, Ballard intends to develop a 50 kW PEM fuel cell small enough and cheaply enough to make the internal combustion engine a thing of the past. Ballard has announced that it is now building a plant to produce fuel cells for as many as 300,000 vehicles a year. The fuel-cell size and cost problems are not entirely resolved, but the technology is developing rapidly and competitively. General Motors Corp. has developed a prototype sedan powered by fuel cells four times as efficient as their 2000 gasoline engines, and GM's vice president of research and development, Larry Burns, has said that hydrogen is the fuel of the future.

Pressure to perfect fuel-cell technology is accelerated by such things as the California law requiring 2% of automobiles sold in that state by 2003 to generate no emissions at all (an example of how a governmental entity can behave proactively for the integration of environmental concerns into industrial practices). Electric cars, which could meet this emission-free criterion as well as fuel-cell operated cars, pose problems (such as their short range) that a fuel-cell car would not have.

Fuel-cell technology is capable of providing a continuous supply of electricity without combustion. It is an environmentally clean, high-efficiency, low-noise energy. Fuel-cell power generation works as shown in Figure 50.

As there is no combustion in a fuel cell, there are no combustion by-products, such as oxides of nitrogen (NO_x), sulphur dioxide (SO_2) or particulate matter—pollutants that are known to harm human health. The fuel cell makes electricity through chemistry, by

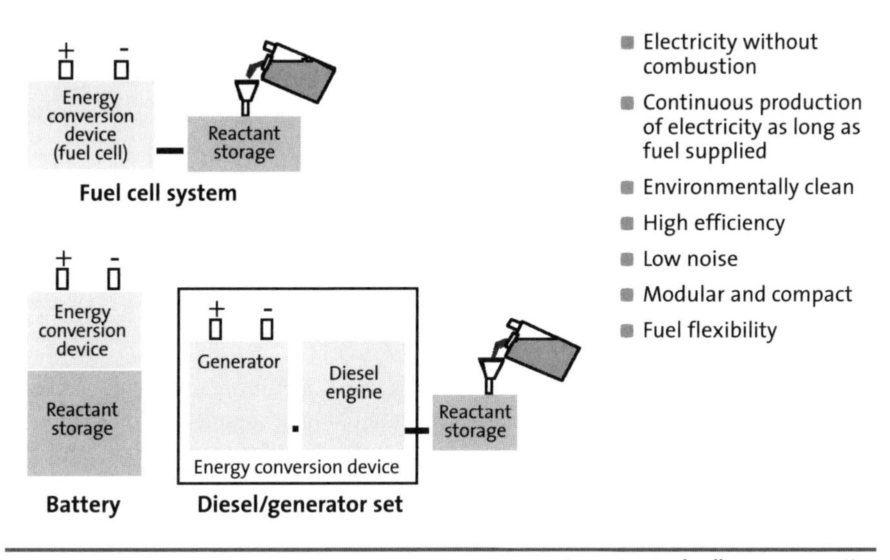

Figure 50 **Fuel-cell power generation**

combining oxygen from the air with hydrogen from natural gas or eventually liquid hydrogen. The only resulting emission is water vapour (Fig. 51).

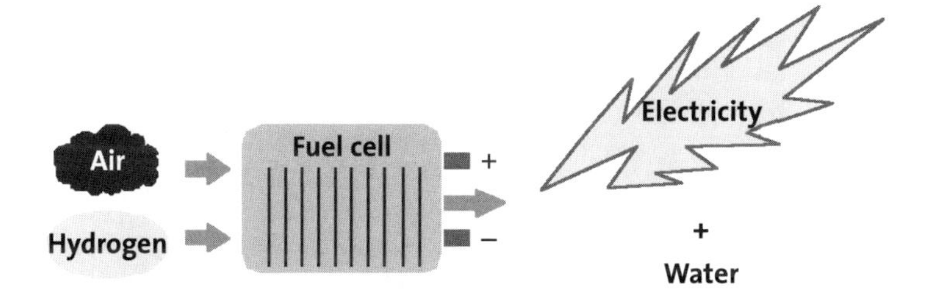

Figure 51 **How a fuel cell works**

Jorge Barrigh (MBA in Finance), who provided much of this information about fuel cells, is with the Ballard Power Systems business development group in Princeton, NJ, in the US. The group's primary objective is the successful introduction and commercialisation of Ballard's stationary PEM fuel-cell power plant product line, and you will be able to follow their progress in the media everywhere.

TECHNOLOGY TOOLS

Innovation born of necessity

Native South African George Couvaras is Managing Director of Sasol Synfuels International (Pty) Ltd, which is responsible for the international commercial application of Sasol's widely acclaimed Slurry Phase Distillate technology for converting natural gas

into high-quality, environmentally friendly diesel. The group has entered into a global joint venture with Chevron for the development, operation and marketing of gas-to-liquids ventures, says Couvaras, who holds an MS degree in chemical engineering from the University of Witwatersrand in South Africa.

Sasol was originally created to produce synthetic fuels from coal to maximise South Africa's self-sufficiency. Given the lack of petroleum resources in South Africa, as well as its political isolation during the apartheid period, it was important for the country to reduce its dependence on imported oil. Sasol thus developed its own technology for transforming natural gas, which was abundant in South Africa, into diesel fuel. Thus the innovation born of necessity.

The oil price increases of the 1970s further stimulated Sasol's growth and innovation. Its move into converting formerly unused or wasted natural gas into a clean diesel fuel placed Sasol at the forefront of environmental innovation. Now, Sasol employs 25,000 people, is South Africa's second largest industrial group, and its operations save the South African economy millions in foreign exchange every year. Currently, the oil industry faces increasing environmental challenges in both its operations and its products. Sasol's gas-to-liquids technology appears to establish a new benchmark for the industry.

George Couvaras came to Paris in 1999 to present the environmental case for Sasol's technology.

INNOVATION BORN OF NECESSITY

Environment-friendly diesel from natural gas

George Couvaras

There are four key issues that make Sasol's derivation of diesel fuel from natural gas of interest environmentally:

1. Natural gas was/is an under-utilised resource; plus there are 1,000 times more natural gas reserves worldwide than oil reserves.

2. Generally, extraction of oil is accompanied by flaring of unwanted or unused natural gas. The volume of natural gas flared worldwide is equivalent to three times the total consumption of gas in, for example, France. Such flaring directly creates CO_2, which leads to global warming. There is an increasing push for a worldwide ban on flaring.

3. Gasoline delivers approximately 20%–25% efficiency in vehicles, whereas diesel delivers approximately 40%–50% efficiency; therefore, use of diesel rather than gasoline lowers CO_2 emissions. In other words, for the same amount of CO_2 emitted, diesel vehicles are able to extract almost double the amount of work of gasoline vehicles.

4. Normal diesel fuels have other noxious exhaust emissions, notably particulate matter and oxides of nitrogen (NO_x). Therefore, the future of diesel fuels is threatened by their inability to meet new emissions standards. Further reductions can be achieved only by exhaust gas after-treatment. Catalytic after-treatment systems are poisoned by sulphur and require less than 50 ppm fuel sulphur. A large investment is required to meet this requirement on conventional diesel fuel.

Thus, out of necessity, Sasol Ltd of South Africa developed a new process, the Sasol Slurry Phase Distillate Process for converting natural gas to diesel fuel with

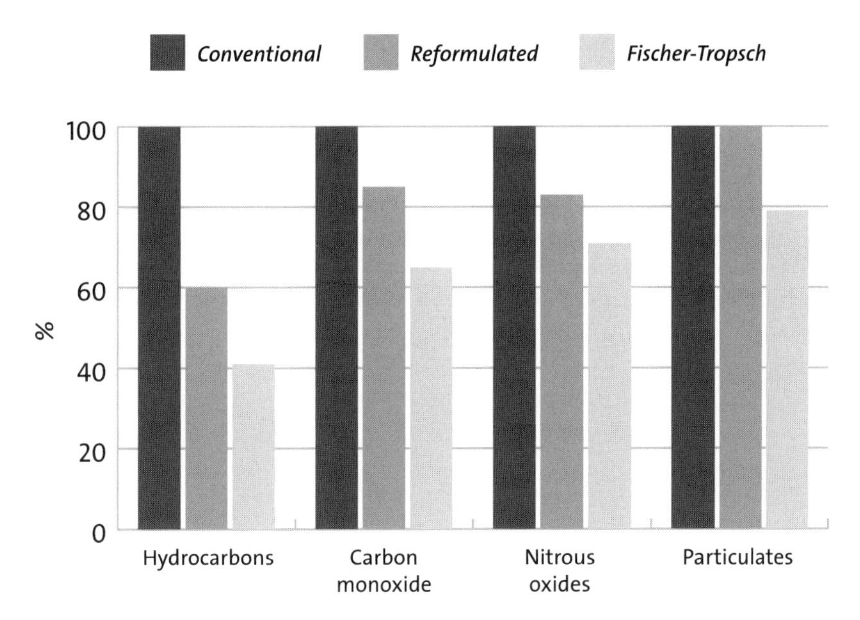

Figure 52 **Fischer-Tropsch fuel lowers all regulated emissions.**

Source: based on engine tests carried out by South West Research Institute, San Antonio, TX

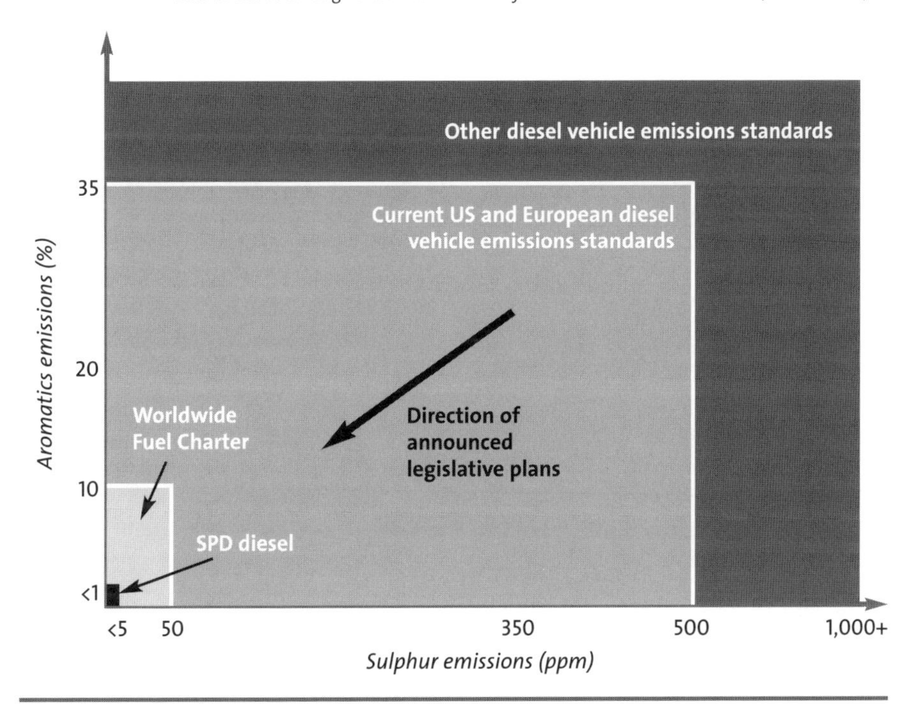

Figure 53 **Sasol Slurry Phase Distillate (SPD) process diesel: setting new fuel quality standards**

extremely low sulphur content via what is called the Fischer-Tropsch Conversion. The low sulphur content allows for catalytic after-treatment. The emission of particulate matter, associated with normal diesel fuels, can lead to respiratory problems. The efficient combustion of Fischer-Tropsch fuel, as well as the extremely low sulphur content, lead to significant reductions in the amount of particulate matter generated during combustion, and the low aromatic content reduces the toxicity of the particulate matter (Figs. 52 and 53).

The Fischer-Tropsch-created diesel is commercially viable, it is distributed via the existing infrastructure, it can be used in existing engines, and it is ideal for fuel cells.

TECHNOLOGY TOOLS
Innovation born of necessity

Consequences of the Montreal Protocol

The Montreal Protocol for the Protection of the Ozone Layer (1987) is an international agreement calling for signatory countries to eliminate the use of ozone-depleting substances (primarily chlorofluorocarbons [CFCs]). As such, the Montreal Protocol has almost inadvertently stimulated rapid technological innovation—innovation born of necessity.

Principal Technical Adviser and Chief of the Montreal Protocol Unit within the United Nations Development Programme (UNDP), Frank Pinto, established (in 1991) and manages UNDP's ozone-layer programme, which covers 1,014 projects in 63 countries in Africa, Asia, the Middle East, Latin America/Caribbean and the CIS, with budgets of $254 million. These projects will eliminate 31,723 tons of ozone-depleting substances (ODSs). As of the end of 1998, the UNDP had disbursed $120 million and eliminated 11,053 tons of ODSs. Pinto has also advised on climate change negotiations, environmental aspects of Chernobyl and the Kuwait oil fires, and on UNDP's energy sector strategy. Earlier in his career, he was Interregional Energy Adviser at the UN Department of Technical Co-operation for Developing Countries, where he advised on energy economics/planning and policy issues, information systems, conservation issues and institutional management. He designed and implemented projects in 27 countries. He has also directed alternative energy programmes at the OECD (Organisation for Economic Co-operation and Development), the Brookhaven National Laboratory in New York and the World Bank.

Frank Pinto received his degrees in economics and finance from the University of Bombay and the Wharton School of Finance, University of Pennsylvania. He has published widely on economics, development, energy and the environment. He wrote the following paper in 2000 for this book.

THE MARKET FOR ENVIRONMENTAL (OZONE-DEPLETING-SUBSTANCE-FREE) PRODUCTS IN DEVELOPING COUNTRIES UNDER THE MONTREAL PROTOCOL

Frank Pinto

The Vienna Convention (1985) and the Montreal Protocol (1987) on the protection of the global ozone layer commit signatory countries to eliminate or replace ozone-depleting substances (ODSs) within a specified period. A Multilateral Fund was established under the Protocol to help developing countries meet the incremental costs of ODS phase-out required under the Protocol. The Fund is managed by an Executive Committee comprising 14 governments, seven from industrialised and seven from developing countries, assisted by a Secretariat in Montreal. It develops operational guidelines and policies and reviews and approves all project and programme financing.

The activities supported by the Fund are implemented through four agencies (United Nations Development Programme [UNDP], United Nations Environment Programme [UNEP], United Nations Industrial Development Organisation [UNIDO] and the World Bank) based on their comparative advantages, technical competence and in response to government–industry request. While UNEP concentrates on information exchange and training and assists low-ODS-consuming countries to formulate country programmes, UNDP, UNIDO and the World Bank are now primarily involved in development and implementation of technology transfer investment projects in the aerosols, foams, fire-extinguishing (halons), refrigeration, solvents and methyl bromide alternative sectors. As of mid-1999, around US$900 million had been received by the Multilateral Fund and most of it had been allocated to these projects.

The pace of technological innovation has been very high under this programme. During 1992–93, the major focus was on 50% reduced chlorofluorocarbon (CFC) formulations. However, by 1994, chemical alternatives that dispensed with the use of CFCs altogether started appearing. It was also about this time that hydrocarbons as an alternative to CFCs in refrigeration started being used in Germany. Thus, during 1994–96, a wide range of newer technologies emerged. Since some of these newer technologies were either expensive, flammable or had other undesirable properties, research continued into more environmentally sustainable options, and since 1997 there has been an increase in water-based and other environmentally safer technologies. This very rapid technological pace resulted in a continuous adaptation of new equipment and products to take advantage of these developments. It also meant that 'experts' had to continuously update themselves on new developments to provide recipient enterprises with the latest possible information and options.

This rapid pace of technological innovation was driven by two factors: first, the banning of the use of CFCs in industrialised countries as of 1 January 1996 except for essential uses; and the bans effective on that same date on manufactured product imports from developing countries that contained CFCs. Thus the more industrially advanced developing countries which depended to a large extent on such exports (e.g. Malaysia, Thailand) suddenly found their export markets lost unless they rapidly adjusted their production processes to avoid the use of CFCs. This resulted in a huge demand on the Multilateral Fund to help these countries and enterprises adjust as quickly as possible.

On the product supplier side, while during 1991–94 there were relatively few equipment suppliers and the cost of non-CFC equipment was quite high, during 1995–97 there was a rapid increase in the number of such suppliers. Suppliers in most countries (e.g. Germany, Japan, UK, USA) decided to adjust independently. However, several large equipment suppliers in Italy decided that this would be a lucrative market in the future and invested heavily in the design of new and improved equipment. As a result, these Italian companies won the lion's share of equipment orders in the foam sector and partly in the refrigeration sector and became the principal suppliers for three agencies—UNDP, UNIDO and the World Bank—by providing top-notch equipment at the lowest prices. The large number of orders placed enabled them to reduce costs, thereby putting further pressure on their competitors. Other suppliers did, however, manage to retain 'niche' markets—such as Danish machines for refrigerant recovery and recycling.

This brings us to what can be expected over the next five years. Most of the larger enterprises in developing countries have completed their conversion away from CFCs and other ozone-depleting substances. What are left are mid-size enterprises and small-scale units, most of which usually export a much smaller proportion of their production as compared to larger enterprises. The small size also means that the cost-effectiveness per kg of CFCs eliminated will also not be as favourable as in larger enterprises.

The challenge for equipment suppliers—both international and domestic—will be to try to develop lower-tech low-cost equipment that can be afforded by these units. This is already being done in several instances. As agencies try to cover the remaining enterprises in each country, greater use will be made of both umbrella projects as well as sector-wide approaches to facilitate implementation and secure economies of scale in orders expected for hundreds of lower-tech equipment units. And, when this is completed, the Montreal Protocol will have succeeded in its objective of assisting developing countries to eliminate use of harmful ozone-depleting substances which will hopefully result in the ozone layer slowly returning back to full strength.

Part 4

Civic actions for change

Civic actions for change, as opposed to the partnerships for change in Part 2, have the potential for changing society's power relations. This distinction is not always easy to make, but examples can illustrate the point. While government-based provision of information can improve the awareness and participation of the public, the active involvement of non-governmental organisations, citizens or citizen groups is necessary to use or develop information in mobilising civil society to challenge the status quo. Such issues as unacceptable industrial emissions, or the choice of location for a potentially offensive installation, or even the choice of what type of installation, system or policy (more efficient rather than less so) invite civic actions for change. We leave it to the reader or to history to decide if any of the following examples are significantly changing the shape of democracies.

PARTY POLITICS

Political parties have traditionally been the most common way of advocating and advancing various causes or sweeping policies that concern whole countries, or, in some cases, the whole of humanity. Green parties have existed in Europe for 20 years, and the author of the following piece began working on policy development within the German Green Party in the early 1980s. Initially, green party policies were not directly concerned with industrial or business practices, but now they are.

Since 1996, Ralf Fücks has been Executive Director of the Heinrich Böll Foundation, headquartered in Berlin. The Foundation is associated with the German Green Party, and Fücks's work there involves sustainable development, the future of employment and international policy. The Foundation works on projects in 56 countries, with special emphasis on Latin America and the Caribbean, South-East Asia and Africa.

Fücks has been a university lecturer, magazine editor, author and journalist, serving as Green Party co-chair and spokesperson from 1989–91. He has been engaged in ecology–economy affairs since the early 1980s and, unlike some of the party purists, strongly advocated transforming the Greens into a government party. He himself occupied several elected posts in Bremen, during which time the city's 'Sustainable Bremen' masterplan was developed, implementing ecological targets in economic, technology, energy, traffic and land-use policy. Fücks and others in the German Greens have ridden the ups and downs of their party's participation in a coalition government with the German Social Democratic Party. The German Parliament's approval of increased eco-taxes on gasoline and electricity during 1999 gave an important victory to the Greens: energy taxes rise annually through 2004.

Two texts follow. The first deals with green ideology and the second with the 'end of ideology'.

WAYS OUT OF THE GROWTH TRAP

Ralf Fücks

> It is scarcely necessary to remark that a stationary condition of capital and population implies no stationary state of human improvement. There would be as much scope as ever for all kinds of mental culture, and moral and social progress; as much room for improving the Art of Living, and much more likelihood of its being improved, when minds ceased to be engrossed by the Art of Getting On (John Stuart Mill [Mill 1848: IV, ch. 6]).

Limits to growth: a brief review

Twenty-five years ago, the first Club of Rome report was published with a title that has since acquired an historical ring: *The Limits to Growth* (Meadows *et al.* 1972). This trailblazing work catapulted a warning about the destruction of the natural foundations of life, as well as a call for a fundamental about-face in politics, industry and lifestyle, directly into social discourse. The finiteness of natural resources and the limited load capacity of the environment were described as nature's barriers to industrial expansion and population growth, whose violation would be at the price of ecological catastrophes. Similarly, the link formerly discerned between economic growth (measured by the gross national product [GNP]) and social welfare was turned on its head: the external costs (ecological, health, social) of growth were shown to already exceed the nominal growth in national income.

The conclusion drawn from this was that **less is more**. In the founding days of the German Green Party, election campaigns were fought on the strength of this motto. Today, the idea would scarcely occur to the party, and would in any case be rejected if only to avoid the appearance of being an accomplice to cuts in social benefits.

Twenty years after the report was published, the ecological theme could look back on a swift international ascent. The Rio Earth Summit raised the concept of sustainable development to the rank of a binding model even if the participating governments accepted no binding obligations. If 'limits to growth' was the starting point for the environmental movement, the emphasis has shifted today. Under pressure from environmentalists, the industrial societies of the North invested massive sums in filter- and sewage-plant and waste-disposal systems—with the result that, in Germany, the air has in fact become less polluted, the rivers cleaner, and the avalanche of garbage channelled into an ordered flow. Industry, meanwhile, has discovered the potential for economic savings that lies in more efficient techniques and recycling processes. This in turn has led to a partial decoupling of economic growth and resource utilisation. Exploration for economically viable raw materials (in particular, fossil energy sources and metals) has grown more rapidly than the consumption of these substances, which means that the threatened collapse of industrial civilisation because of a lack of natural resources has once again receded into the future.

Sustainable growth: the Utopian phantom

This 'all-clear' led to the birth of the Utopia of **sustainable growth**—the expectation that permanent technical innovation will minimise resource consumption and pollutant emissions to such a degree that new markets can be created and higher employment levels maintained while reducing resource use. In other words: ecologically correct growth as the phantom solution to all our needs. The happy message reads: we can escape the eco-crisis simply by waving the magic wand supplied by the scientific-technological revolution ('Factor 4'), and at the same time rescue the welfare state (which is based on the distribution of permanently increasing output levels) and the hedonist lifestyle of the urban middle class.

Franz Alt (German television journalist and author known for his optimistic view on the economical and cultural benefits of sustainable development) hit the nail on the head with his description of the harmonious link between ecological turnaround and economic offensive in international industrial competition:

> The first industrial countries to grasp that in the 21st century, deriving energy from sun, wind, water and biomass will be at least as important as automobile production in the 20th century, will be positioned to transact business with the whole world, to have no unemployment problems, to secure their national production sites, and to do a service to nature. So who's going to be the first?

This unofficial alliance for ecological modernisation allows for a wide range of actors: for enlightened entrepreneurs and trade unionists alike, for engineers, ecological farmers and trend-setters, for Green Party members as well as the forward-looking wings of their political rivals. In a prospective process of ecological restructuring fuelled by tax reform, equally important and promising elements would be solar technology and power–heat coupling, low-fuel cars and hydrogen-powered aircraft, energy-efficient household appliances and energy-saving buildings, miniature technology and recycling circuits, renewable resources and soft biotechnology, as well as the shifting to data highways of communications.

Eco-innovation: the semi-ecological truth

Only a fool would turn down the chance to link ecological necessity with economic expedience. The Greens and the environmental movement might wait a long time for another strategic constellation that allows them to join in delivering the most promising suggestions for Germany's future as an industrial location. Nevertheless, doubts remain about the feasibility of sustainable growth as the lowest common denominator between ecology and economics.

While landscape destruction, reduction in the variety of species, deforestation, depletion of once-fertile soil, over-fishing and marine pollution continue unhindered, the efficiency gains in the power sector or automobile industry have been swallowed up by a permanent increase in output quantities. In the highly developed states in Germany, eco-technical advances have effected only a levelling-off in the already excessive levels of water and power consumption and nitrogen and CO_2 emissions, while in the rapidly expanding new industrial regions of southern Germany, the burden on the environment is growing at an alarming pace.

It is safe to say that, because of the consumption of natural capital it entails, our mode of production and life cannot function as a model for the remaining 80% of humankind unless the natural conditions of life are to be irreversibly ruined. It is also true that, over the coming decades, resource consumption and emissions in the industrial northern metropolises must be reduced by 80%–90% in order to allow the communities in the south the necessary scope for economic and social development. Thus, 'ecological modernisation' and 'exit from the growth spiral' are not alternative options, but different sides of the same coin. It is not only a question of more environment-friendly production facilities, products and infrastructure, but of an absolute reduction in material production and consumption.

The return of growth mania

In times of worsening unemployment and sinking income, there is something apparently frivolous about criticism of economic growth, which tends to be written off as a faddish luxury indulged in by educated professionals with guaranteed pensions. Growth sceptics stand largely on their own—from employers' associations to left-wing/alternative professors of economics, from the Free Democrat Party to the Social Democrat Party, there is unanimous agreement about the need to boost economic growth as a way of mastering social and fiscal problems. Their arguments primarily deal with the question of how: by progressive deregulation, a policy of financial restraint and the reduction of enterprises' operating costs, or through a policy of forced relaxation of interest rates, loan-financed state investment programmes and the boosting of households' purchasing power. It is the old sparring match between adherents of the supply-and-demand theory, monetarists and Keynesians.

Disciples of growth ideology of the 1960s and '70s become all the more fervent as their model falls apart in face of the realities. As everybody is aware, even a more than 2% increase in GNP scarcely benefits the labour market, or public finances either—meaning anticipated tax revenues have to be reduced. Anyone who banks on growth to get unemployment under control and salvage the welfare state is on unsteady ground—and risks the irreversible ecological collapse of our planet.

Sinking growth rates: an economic trend

At the same time, there are good reasons—economic ones included—to assume the high-growth era is over for the foreseeable future in traditional industrial societies. Technical progress is concentrated on process, as opposed to product, innovation. Its main effect is rationalised goods production and services, but not the formation of new markets for consumer goods. This is equally true of modern information and communications techniques. Fundamental innovations that might justify a new cycle of domestic demand (analogous with the invention of the automobile, electrical domestic appliances, or consumer electronics) are lacking.

Moreover, the increase in capital intensity is accompanied not only by a rise in the cost of every new industrial and skilled-services workplace but also, the more developed a national economy is, by progressively rising growth-incurred social costs such as, for example, public-financed research and development, academic training, information and traffic infrastructure, investment and export promotion. Every single percentage of growth thus requires a growing volume of advance state payments, while the tax revenue per growth unit sinks. This also entails a drop in the price-to-performance ratio of public investment programmes, accompanied by an alarming increase in public debt levels.

Politics in the age of less

When critics of growth mania are confronted with the argument that labour markets will collapse without growth, social security systems will no longer be financially viable, and public finances will veer out of control, then the opposite is in fact true: none of the major social issues can be solved with growth policies.

Politicians with foresight must adjust to the idea of sinking economic growth rates. We have left behind the times when the fight for a portion of distributed wealth could be neutralised with the motto 'more for all', and a special finance programme and a new public service department could be set up for each problem group. As a political instruction manual, the popular 'both . . . and' has outserved its useful life—in the future, decisions will have to be made on the principle of 'either . . . or' and clear priorities must be set.

If growth therapy no longer functions, social crises must be treated with social innovations. The question of the future shape of labour and social security, of the composition of a tax system and public finance policies that cannot rely on permanent economic growth, is a revolutionary challenge to academic and social creativity. Some elements of this reinvention of society for the ecological age have been under practical trial for some time now—in models such as agricultural co-operatives combining producers with consumers, ecological city quarters, the 28-hour working week collective agreement for the Volkswagen workforce (more leisure time for less income), and in countless social and cultural initiatives that mobilise voluntary commitment in a citizens' society.

Moderation as an ecological imperative

If sustainable development is to be more than just cosmetic surgery, then we need both of the following: **ecological-technical efficiency** and the **willingness to be moderate**, to restrain our own material expectations.

Even low-fuel cars and blanket train networks will not make environment-compatible the current level of mobility in the industrial metropolises—and a true turnabout in traffic policy entails not only less long-distance haulage, but also less recreational traffic and flights. The permanently rising need for residential building land, and the enduring tendency towards detached family houses is a declaration of war on environmental protection, and keeps power consumption high. The ecological motto is: **move closer together**. Anybody interested in environmental issues knows that the current levels of meat consumption are no less reconcilable with animal-friendly farming and ecological agricultural methods than are the low-price foodstuffs in supermarkets—but who is willing to break the news to the masses who see steak on their plate as a symbol of prosperity?

However one chooses to phrase the message that 'less' (paid employment and income) also includes a chance for 'more' (leisure, creativity, social life), there still remains an element of sacrificing current ways of life and consumer expectation—and this applies not only to the rich. But in traditional left-wing/alternative politics, the notion of doing without was tainted when real income began to drop and social services were pruned in the course of globalisation and the financial distress of public authorities. A policy of ecological moderation currently runs the risk of appearing to be an ideological heightening of the actual reduction in mass income taking place at present; but the policy need not succumb to this.

A new politics of social justice

What we need to teach ourselves is a new politics of social justice directed not at distributing growth and defending existing assets but at fairly distributing the 'less'. At the same time, these policies must offer guarantees against a social descent into poverty and insecurity, and open up new possibilities of the good life, ranging from the advantages of public transport over private to the recreational possibilities in local unspoiled landscapes up to the access to education and culture for all. If we want to build bridges to a life not geared towards maximising individual consumption, then we have to prevent public welfare and communal facilities from being increasingly decimated in the course of public-spending cuts—whether this applies to state schooling, citizen-friendly municipal institutions or subsidies granted to child–parent groups, sports associations or cultural initiatives. Taxation and budgetary policies that add to the private wealth of a social minority at the cost of public impoverishment are violations of sustainable development.

Civil society and leisure time

Obviously, the marketplace and the state will remain the central planes on which future societies act and regulate. Yet it is foreseeable that a social order constituted on a sustainable basis will allot more weight to non-commercial initiatives and citizen-organised bodies—to the essential domain of civil society, in other words.

It is not a question of shutting down the monetary economy, marketplace and economic competition, but of limiting their sphere of influence and of re-integrating the economy into society, instead of subjugating society as a whole to economic principles: the market economy, then, as opposed to market society. If the other extreme—an increasing public ownership of society—is to be avoided, then the intermediary area of direct social activity conducted outside the confines of the market (salaried work) and the state (bureaucracy) must be expanded.

Necessary for this is an increase in **free, self-determined time** for one and all. Ecological politics seeking a way out of the growth trap must enter into **a strategic alliance with policies seeking a reduction in working time**, whether the model foreseen is the continuous, productivity-based reduction of weekly working hours, or variable annual and lifetime work durations, grants for time off for child-raising or further education, or part-time work with a guaranteed level of social security.

Our common aim should in future be to steer technical progress in two directions: on one side into the ecological-technical efficiency revolution—that is, into the development of environment-friendly, resource-efficient products and processes—and on the other side into the continuous reduction of socially necessary work (Marx), instead of into the unlimited expansion of apparatus and goods production. Furthermore, this strategy will reinforce the foundations of civil society, because it offers improved opportunities for the maximum number of citizens to participate in the public affairs of the community.

THE GERMAN GREENS AND THE END OF IDEOLOGY

Ralf Fücks

The following are excerpts from a 1999 talk by Ralf Fücks at the American Enterprise Institute in Washington, DC.

. . . Engaging in politics is an open process, an act of freedom. And of course this simple truth applies to Green politics as well . . . In their early years, the Greens were above all a negative coalition: against nuclear energy, against the arms race, against the military-industrial complex, against risk technologies, against subjecting minorities to discrimination. The positive consensus was quite abstract—a canon of fundamental human and political values: peace with nature, responsibility for the coming generations, activism for democracy, human rights, social justice, and non-violence.

. . . In the early eighties, the Green mainstream was not very interested in pursuing a homogeneous strategy. It was an era of 'politics in the first person', driven by personal convictions and values that seemed self-evident.

. . . [Now] Green ideas and concepts have moved from the fringes to the centre of society and politics. In the meantime, issues such as sustainable development, gender democracy, making state politics accessible for citizen participation, and giving priority to non-violent conflict resolution, have entered into the mainstream—no matter how far away we still may be from fundamental changes in economic and international policies.

. . . [The Greens] have become professional in a triple sense:

- First, by practising politics as a specialised profession
- Second, by gaining concrete qualifications in more or less all social and political issues while developing concrete alternatives for reform in the energy sector, the tax system, social security, the educational system and public transport

- Third, by accepting the different roles played by political parties and non-governmental organisations, by parliamentary policy and extra-parliamentary action

. . . Today, what specific political corner do the Greens and their programme represent?

From the outset, the Greens brought a new element into the political landscape. Novel aspects included the repudiation of permanent economic growth as the basis for the welfare state, the government-critical orientation toward self-initiative, self-administration and local networks, as well as the principle of 'generational equality', which manifested itself in the demand for ecological responsibility for the future long before the current debate about retirement benefits: 'We have merely borrowed the Earth from our children.'

The Greens were the party of 'civil society' long before that term ascended to the top of the political charts. We want to encourage social activism: by way of complementary private–state funding models, tax incentives for foundations and volunteer work, and even granting retirement benefits for civic activities. Before long, we should again address the issue of providing a tax-financed basic income for all, linked to the concept of having all citizens engaged in charitable activities. For reasons of both justice and funding, 'public money' and 'public works' are terms that go hand in hand.

COMMUNITY-BASED CURRENCY AND EXCHANGE

Richard Douthwaite is interested in identifying the unsustainable characteristics of our economic system so that they can be changed. In 1998–99 he worked on a European Union-funded project to establish experimental community currency systems in Scotland, the West of Ireland, Amsterdam and Madrid. *The Ecology of Money* (1999), which proposes a radical restructuring of national and international money systems, grew out of this experience and his work as economic adviser to the Global Commons Institute, which is concerned with halting human-induced climate change.

His first book was *The Growth Illusion: How Economic Growth Enriched the Few, Impoverished the Many and Endangered the Planet* (1992). His second, *Short Circuit: Strengthening Local Economies for Security in an Unstable World* (1996) gives examples of currency, banking, energy and food production systems that communities can use to make themselves less dependent on the world economy. These ideas are discussed in the following piece.

THE SHORT-CIRCUIT APPROACH

Richard Douthwaite

Not much more than a century ago in Western Europe and the US—even more recently elsewhere—almost everything that people used and consumed came from their immediate area. Their food came from their own land or from a neighbour's. Their timber and fuel came from a local wood. Many manufactured goods were local, too, as shown by this 1912 account of how life in Farnham in Surrey had changed in the previous 50 years:

> It is really surprising how few were the materials, or even the finished goods, imported at that time [the 1850s]. Clothing stuffs and metals were the chief of them. Of course the grocers (not 'provision merchants' then) did their small trade in sugar and coffee, and tea and spices; there was a tinware shop, an ironmonger's, a wine-merchant's; and all these were necessarily supplied from outside. But, on the other hand, no foreign meat or flour, or hay or straw or timber, found their way into the town, and comparatively few manufactured products from other parts of England. Carpenters still used the oak and ash and elm of the neighbourhood, sawn out for them by the local sawyers: the wheelwright, because iron was costly, mounted his cartwheels on huge axles fashioned by himself out of the hardest beech; the smith, shoeing horses or putting tires on wheels, first made the necessary nails for himself, hammering them out on his own anvil. So, too, with many other things. Boots, brushes, earthenware, butter and lard, candles, bricks—they were all of local make; cheese was brought back from Weyhill Fair in the wagons which had carried down the hops; in short, to an extent now hard to realise, the town was independent of commerce as we know it now, and looked to the farms and the forests and the claypits and the coppices of the neighbourhood for its supplies. A leisurely yet steady traffic in rural produce therefore passed along its streets, because it was the life-centre, the heart, of its own countryside (Bourne 1969: 103).

Today, of course, all that has changed utterly, and the food, fuel and manufactured goods we need for our survival are transported over very long distances to get to us. As a famous German study (Böge 1993) showed, even the components required to produce and deliver a pot of strawberry yoghurt had to travel a total of 2,183 miles. The huge increase in transportation has, perhaps, led to a proportionate increase in the range of product choices and the abundance available to people like us with enough money to benefit. However, our gain has been bought at great potential cost. The longer supply chains we rely on are less sustainable than those they replaced because of the amount of fossil energy required to keep them running. This in turn means that our survival depends on the continued availability of oil and gas at a price we can afford, something that is looking increasingly uncertain. The International Energy Agency accepts that world oil output will peak within the next ten years and that, after 2008, over half the declining output will be controlled by five OPEC countries (Organisation of Petroleum Exporting Countries) in the Middle East, giving them enormous power over its price (IEA 1998).

The longer supply chains present another serious problem, too—they are much more liable to be captured by large corporations who then use their power over them to make larger profits for themselves. If 98% of the retail food trade in a country is controlled by just four companies, as is the case in Norway (*Irish Times* 1999), then food producers are in a very weak position. Each manufacturer needs to sell to each retail chain to survive but the retail chains do not necessarily need each manufacturer and will drive its prices down.

This process widens the margin the distribution system can take. In Ireland in 1973, when there were 35,700 family farms producing pigs and bacon factories were dotted all over the country, half, on average, of the price the consumer paid for a pork or bacon product went back to the farm.[23] By 1996, however, although the total number of pigs reared in the country had more than doubled, there were only six bacon factories of any size, three of which were controlled by one company. As a result, just over one-fifth of the price of a pack of bacon was getting back to the farm.

Similar changes have occurred almost everywhere. In the United States between 1980 and 1987, for instance, the amount the farmer received for his contribution to a box of cornflakes fell by one-third while the price of the box to the consumer went up by the same proportion (Goering *et al.* 1993: 34). In the Federal Republic of Germany in the 1950s, three-quarters of all spending on food went back to the farm. Thirty years later, the proportion had dropped to one-fifth (Greenpeace; see IEA 1998). And it is not just agriculture that has been squeezed either.

Very much the same has happened in the financial sector where the rate of interest offered to the small saver is now a much tinier proportion of the rate at which he or she could borrow than it was even 25 years ago. Similarly, firms that manufacture relatively simple unbranded goods have been getting a much smaller fraction of the retail price too.

These growing gaps between bid and offer prices present a wonderful opportunity for anyone hoping to achieve a healthier, less risky balance between the things the people of an area produce and do for themselves, and those that outsiders produce or do for them. Indeed, exploiting these gaps is a key part of the

23 *Business and Finance*, 27 August 1992.

short-circuit approach. If we can grow food and bypass the supermarket system to sell it to our neighbours, then not only should the producers be able to get a larger proportion of the price the consumers pay, but the food chain will become less energy-intensive and thus more sustainable. Or, again, if we can lend money to each other without having to go through the commercial banks, both borrowers and lenders should be better off and less exposed to the effects of fluctuations on international financial markets.

There's rather more to the short-circuit idea than simply cutting out the middleman, though. Essentially, it is a systematic way of equipping one's local economy so that a much wider range of activities can be carried on than would be possible if external prices, investors and systems were the sole arbiters of the range of goods and services to be produced. Credit unions and box schemes are just examples of the required equipment.

The fundamental short-circuit step is equivalent to erecting a polytunnel to grow fruit, vegetables and flowers that would be stunted or killed if they were exposed to the harsh, cold winds outside. In other words, by establishing a local currency system and a local bank, the short-circuit approach seeks to create a favourable local financial micro-climate in which many more projects can blossom than would otherwise be the case.

'Why a local currency?', you might ask. The answer is that circumstances often arise in which people living in the same area would like to be able to buy and sell to each other but are unable to do so because they don't have enough money. The money they require doesn't have to be in the form of national currency, however. As the hundreds of Local Exchange Trading Systems (LETS)[24] that have been set up in the past ten years have shown, anything will function as money so long as it provides a scale by which the value of the work done by the person who comes to paint my house can be compared with the value of the work I do for them or for another of my neighbours. Once we have established such a scale of value, we can make the level of activity in our local economy less dependent on the inflow and outflow of money created elsewhere.

What's more, a local currency gives firms that can quote their prices in it a significant trading advantage over firms that can't. This is because local money will always be easier to acquire than money that has to be earned—or borrowed—from people outside the area.

Having a local bank or investment organisation brings a lot of advantages, too, quite apart from the one already mentioned—its ability to give both borrower and lender a better deal. This is because it can take factors other than the rate of interest into account when making its loans. When people are investing outside the areas in which they live they can only be interested in one thing—the rate of return they get on their money. All the other income streams that their investment starts—payments to workers and suppliers, for example—are seen as reducing their profits, and every effort is therefore made to minimise them.

24 www.sussex.ac.uk/Units/gec

If someone invests in a project in their own community, however, there are many ways in which they can get a return on their money quite apart from the interest they receive. Indeed, these non-interest returns might be so important that those financing the project might be prepared to charge no interest at all and even contribute to an annual loss in order to be sure it goes ahead. This might be because the project will provide employment for themselves or their children. Or because it will increase incomes in the area and help their existing business do better. Or because it will cut unemployment, thus reducing family breakdown and crime.

Community investment projects are therefore very different animals from those run by outside investors. For one thing, they seek to maximise the total incomes the project generates in the community, not just the profit element. So, far from seeing the wage bill as a cost to be minimised, they regard it as one of the project's major gains. Attitudes to work are different, too. Whereas outside investors seek to de-skill work within the factory so that they can hire the cheapest possible labour, a community company, particularly a workers' co-operative, would want the work to be organised so that those doing it find it interesting and fulfilling.

Outside investors also have very short time-horizons for their projects, wanting to earn their capital back in three or four years. After that, if necessary, they can close the plant and move on. Communities, on the other hand, need long-term incomes for long-term projects such as raising children, and a community-owned factory would want to produce for a safe, stable market, most probably in its own area, rather than the market with the highest immediate rate of return. Similarly, while outside investors merely ensure that a plant's effluent, noise and emission levels stay within the law because anything better would cost them money, a community company is likely to work to much higher standards to avoid fouling its own nest.

In a nutshell, the only way that a genuinely holistic assessment can be made of the full costs and benefits of a proposed investment to the entire community is if the whole community, or an organisation representing it, carries it out. The local bank or investment organisation should be such a representative body. It should also be structured so that it can provide things that are potentially much more valuable than equity capital and loans—help and advice. If you are lending to your neighbour, you cannot behave like a commercial bank and tell him or her 'If your project fails and you can't repay, I'll seize your house and sell it to recover the money.' Neighbours shouldn't treat each other like that. Instead, you have to get involved in the project while it is still in the planning stage and make sure that it is properly researched, resourced and structured. After the loan is made, you have to stay involved, offering advice and practical help, for as long as is required.

This formula worked very well for what is perhaps the best-known community investment agency in the world, the Caja Laboral Popular, the community-owned bank at the heart of the Mondragon co-operatives in the Basque country in northern Spain. There, if a group of people want to set up a new co-op, they choose one of their numbers to work full-time with the bank, on bank rates of pay, to prepare the plans. They work with 'godfathers'—bank employees who have helped set up many businesses in the past. When the godfather thinks that the project is ready, he recommends it to the board for financing and then becomes a director of the new co-op for as long as the bank has money invested in it. As a result, it is almost unknown for these start-ups to fail.

With these two bits of equipment—a local investment organisation and at least one local currency—it becomes very much easier for a community to provide itself with more tools. It can start a farmers' market, for example, and, if one no longer exists, build and equip a small abattoir so that local cattle and sheep don't have to travel long distances to be killed and then get lost in the mainstream food chain. Or it could start a dairy that makes butter, yoghurt and cheese.

In the energy area, the bank could help local farmers to start planting quick-growing trees such as willow to supply woodchips to another bank project, a combined heat and power station, which would not only generate electricity but also supply hot water for central heating to shops, offices and homes in its immediate area. Many power stations of this type have been in operation for years in Finland and Austria. Or a local credit union could give loans of up to £5,000 each to 100 families to enable them to set up a wind-power co-operative, just as dozens of local savings banks have already done in Denmark, and with great success.

The short-circuit approach therefore involves mobilising local capital and other local resources to meet local needs in order to make each area much more self-reliant and sustainable. It is in direct opposition to the idea that local capital should be invested wherever in the world it can reap the highest financial return and that other local resources should be used to make products and to provide services such as tourism for customers in markets far away.

This latter idea is not wrong. It brought appreciable benefits in its day but, as so often happens in human affairs, it has been taken to such extremes that it is now dangerous and out of date. Its over-use has created the situation in which very few communities are able to do anything significant for themselves. If they are not to remain that way, and thus completely exposed to the risks presented by an increasingly unstable global economy, each community needs to recover the ability it once had to meet its own basic needs. It can do this best by developing its own independent monetary and financial structures. And that is what the short-circuit approach is all about.

THE MEDIA

The power of the press to ignore or draw attention to issues cannot be overestimated. The environment, beginning around Earth Day in the 1970s, became a hot subject, but in the 1990s it was dropped for a period of time, perhaps because it wasn't novel, perhaps because it was boring, or because it was too apocalyptic. In 1999, the subject began to reappear in the major press, perhaps to coincide with the millennium, perhaps because there are genuine changes afoot, which makes the subject new again.[25]

Todd Gitlin has much to say about the difficulty of finding serious environmental coverage in the media. He is the author of seven books, including *The Twilight of Common Dreams*, a novel, *Sacrifice*, and a new one entitled *Media Unlimited: Life in the Torrent of Image and Sound*. He is also a lecturer, a columnist and a professor at New York University in communication, journalism and sociology.

Following are excerpts from an interview with Gitlin which appeared in the *Wild Duck Review* 5.1 (Winter 1999; Nevada City, CA), in which he warns: 'Don't leave environmentalism to the pure-of-heart.' This, we agree, is worse than leaving politics to the politicians. Next is a follow-up to the same subject that Gitlin wrote in early 2000 for this book. As a practising journalist, he wrote on the environmental aspects of the December 1999 World Trade Organisation (WTO) meetings in Seattle in *Newsweek International*, *Los Angeles Times*, *Ha'aretz* (Israel) and the Internet magazine *Salon*. He has continued to write in a variety of publications on the protest manifestations surrounding subsequent international meetings.

25 By 2001, 'sustainable development' is the subject of advertising supplements in publications such as *Business Week*.

MEDIA/ENVIRONMENT

Todd Gitlin

. . . Political institutions, including the press, more or less collude in keeping huge questions off the agenda, either because they're blind or because they're wilfully deceptive and exclusionary. Then it takes social movements to bring these questions to the fore. Only then do we get a necessary debate. This is the case now with environmentalism.

. . . To me, the question is: Where can the debates take place? Obviously, they're not going to take place on ABC TV, and they're not going to go on in the *Wall Street Journal* or the *New York Times.* I'm holding in suspense the question of whether the right ideas exist somewhere on the planet. But surely a great deal is known about strategies for healthy systems, and the various positions ought to contend if there are going to be democratic decisions . . . I have been involved in conversations enough over the last few years to know there are debates, there are proposals, and that these are global. What is frustrating is that you'd have no idea, from moment to moment, as a US citizen, let alone as an activist, of the range of these discussions.

. . . For example, in 1997 I was invited to a conference in Paris called ECO 1997. The approach of this conference was to bring people together who were either business or political powerhouses, and discuss practical schemes in Europe, USA, Japan and South America, schemes that are either in place or contemplated and promising. The emphasis was on pragmatism. Now, I found it extremely interesting, very different from the abstractions so gaily and formulaically tossed around on the left. I was invited simply to ask tough questions of the panellists, as was Mark Hertsgaard, the journalist. The two of us were lobbing questions at EEC figures, the German Environment Minister, the French Environment Minister, Clinton Administration people, executives from Dow Chemical, AT&T, Nissan Motor and so on. I came away somewhat heartened by the knowledge that there were powerful people talking about practicalities—what is actually do-able, on this side of some millennial fantasy of revolution. I don't know if I'd call it a dialogue, but at least there was a respectful expression of positions. Some participants were promoting forms of industrial activity that seemed to be common knowledge to the specialists but were new to me. The so-called eco-efficiency arrangements are not out in the open for public discussion. The people who organised the conference worked hard to get out the word about what they were doing, to get press coverage, but unfortunately without much success. There's a second such conference in Paris this June, ECO 1999, and I hope the media pay more attention, though I am not holding my breath. There's a specific example of a framework for thinking that is constructive—not millennial, but constructive—and is simply unknown to Americans.

Following up on media/environment in early 2000

The showdown at the World Trade Organisation meeting in Seattle in December 1999 sent auspicious signals for environmentalists, promising a revival of eco-activism in long overdue alliance with trade union muscle. But it takes tremendous clarity and work to compensate for the lost opportunities of recent years. On the positive side was the eruption of unexpected numbers in carnivalesque array, co-ordinated across frontiers, militant and non-violent, both sober and playful, and at the same time sensitive to the need to cement alliances with other social forces.

Both the media and political authorities of various persuasions quickly understood that the black-masked anarchists smashing corporate windows constituted a tiny minority of the total demonstration. US President Bill Clinton saw fit to welcome the demonstrators, conducting a respectful argument with them, taking up their cry to place labour rights on the agenda of trade negotiations, suggesting how substantial the protesters were, for they represent two forces (and more) that the left has been hoping to unite for decades now: not only the environmentalists and indigenous groups who have been following out some of the strands of the long-gone 1960s, but also labour, which used to be mightily suspicious of green types, whom they saw as privileged young scamps insufficiently appreciative of people who work for a living. In a phrase, we witnessed the public emergence of a Greenie–Sweeney (president of the United States labour federation AFL–CIO) alliance.

Now comes the moment of pragmatic application. For movement of street spirit does not translate directly into political power. It offers a psychological boost to the hardcore, but such boosts are evanescent. The joy of stopping traffic remains a puny force unless it translates into the labour of everyday recruiting, and this must become a politics that engages the hearts and minds of the majority who fear unbridled corporate power—the real target, albeit frequently misnamed 'globalisation' or blamed on the largely symbolic World Trade Organisation.

On this score, protesters—and not only Americans—will soon have to choose between two distinct approaches to the WTO and what it represents. In Seattle, the distinctions could be fudged, in the interest of bringing together a broad alliance. But a hugely consequential political choice will now have to be made between two mentalities. There are those, essentially nationalist, who want to destroy the WTO because it encroaches on national sovereignty. And there are others, essentially internationalist, who recognise that the environmental and economic changes the world needs require not less supranational government but more. The demand to make the WTO responsive to labour rights in Indonesia, India or China is a demand for more supranational government. Equivalent demands to strengthen other supranational authorities (such as the European Union) in the interest of environmental sustainability will also require global authorities. So, too, the increasingly audible demand to create an international constabulary responsible to the Secretary General of the United Nations and capable of being mobilised to stop mass slaughter in such places as Rwanda, Kosovo and East Timor. Such demands, in fact, add up to a demand for a new 'hyperpuissance' capable of enforcing international law—not the United States, but a more general, more legitimate force.

The spirit of federation is basically right. Although opposed by the governments of many developing nations, the demand for legitimate transnational power is supported by indigenous peoples, and unions of the left, and many environmentalists, who want international regimes to produce not a weakening of human rights, environmental protection and labour protections but a strengthening. Not for them—for us—the 'race to the bottom', the weakening of national controls on polluting, fishing and so forth, but the enforcement of strong regulations against pollution and species depopulation. For there is no alternative to globalisation—of capital, imagery, protest, culture or influence. The questions are rather: Who controls globalisation? Will it be conducive to economic justice? Will it help preserve life on the planet?

ENVIRONMENTAL ACTIVISM

Environmental activism in relation to industry has a long history, but the varieties are proliferating. The following article describing a kind of tense co-operation between environmentalists and the Dow Chemical Company appeared first in the *New York Times* and then in the *International Herald Tribune* on 19 July 1999. It is reprinted here with permission. This sort of industry–environment story is increasingly turning up in major newspapers, just as the kind of action described is proliferating.

A DARING PARTNERSHIP PAYS OFF

Activists help teach Dow Chemical to cut pollution—and costs

Barnaby J. Feder

Midland, Michigan—They all played their classic roles once word got out last November that a Dow Chemical Co. contractor had repeatedly spilled toxic dust intended for Dow's incinerator. Denunciations poured forth from outraged environmentalists such as Diane Hebert, chairwoman of a local group called Environmental Health Watch. 'We are shocked by the carelessness and indifference that allowed these dangerous spills to occur,' Ms Hebert told the Detroit Free Press. 'Dow's negligence has endangered our community's health.'

Dow was a predictable corporate heavyweight, announcing that it was co-operating with an investigation by state regulators and barred from saying much. Local coverage, trotting out old descriptions of the wastes at the plant site, made it appear that Jeffrey Feerer, the environment, health and safety manager for the manufacturing operations here, was minimising the presence of dangerous dioxin by characterising the spills as little more than 'sand and dead microbes'.

But even as the adversaries squared off in public, they quietly continued collaborating on a novel two-year drive to slash toxic chemical emissions from Dow's vast chemical complex on the eastern edge of town. The two sides were out to discover whether environmentalists armed with detailed information about a company's business needs and processes could help it find and carry out profitable ways to cut waste.

It was an uncomfortable alliance, fraught with risks for both sides. Some people at Dow feared that giving some of their harshest critics such an intimate look at operations would blow up into a public relations nightmare or lead to leaks of confidential information to competitors. Ms Hebert and the other environmentalists involved agonised over whether Dow might exploit the exercise to make itself look environmentally responsible without producing meaningful changes.

As the clash in November and previous public battles over a new waste-water treatment project here proved, even the most successful collaboration could not paper over all of the adversaries' long-standing differences. (The investigation into the spill is still pending.) Still, everyone stayed the course. By the time the project was completed on 30 April [1999], Dow was on track to cut production of a list of toxic chemicals selected by the environmentalists by 37% and to reduce the release of the chemicals to the air or water by 43%. To the surprise of both sides, the project had beaten a goal of 35% reductions on both measures, which had seemed a stretch in the beginning.

Moreover, Dow's investment of $3.1 million to make the changes is expected to save it nearly $5.4 million a year and, for some businesses, improve product quality or add to production capacity. Companies and environmental groups have worked together in the past on pollution prevention. But the depth and length of Dow's relationship with the environmental activists are prompting talk of a new era of co-operation.

'This partnership will almost certainly become a model nationally among companies looking to improve the environment and improve their bottom line', said Carol Browner, administrator of the Environmental Protection Agency. The groundwork for the project was laid in 1996 by Dow and the Natural Resources Defence Council, an environmental policy group based in New York. They recruited five long-time local critics of Dow, a step that both sides agreed was crucial to the project's credibility and its impact on local managers. The challenge was to find big reductions that would actually save Dow money—a search for so-called low-hanging fruit.

Dow believed there were few opportunities for reductions because the site had been working on pollution controls for two decades. The environmentalists, assuming that businesses were adept at spotting profit-enhancing environmental projects, also had their doubts. But Dow managers and their new environmentalist advisers quickly found a startling array of prospects that had been routinely overlooked. Dow, for example, discovered that introducing new chemical catalysts and making other innovations in the manufacture of resins, combined with modifications in the processing equipment, eliminated a nasty by-product, formaldehyde-laced tars. The one-time cost of the project was $330,000, while the savings at Dow's waste-treatment plant totalled $3.3 million annually.

Dow managers became especially intrigued by the money-saving moves turned up by Bill Bilkovitch, a tireless environmental engineering expert based in Tallahassee, Florida, who was hired by Dow and spent more than two years at the site. 'Many of these projects were too small to be seen as real business opportunities', said Linda Greer, an environmental toxicologist with the Natural Resources Defence Council, who worked with Dow to recruit the local environmentalists and organise the project. As surprising as the savings were, the most provocative aspect of the project was the crucial role the environmentalists played and Dow's willingness to lower its guard. 'It was a heroic decision to bring in their most vocal critics', Ms Greer said.

Dow let the environmentalists choose which toxic chemicals to go after. It accepted their list of 26 without argument, even though many were selected mainly because they can produce dioxin when burned. That required stifling any impulse to defend the company's long-standing position that critics exaggerate dioxin's health risks and industry's role in creating dioxin. The company also accepted the aggressive reduction target of 35% and, with many qualms, the environmentalists' refusal to sign any confidentiality agreements.

Perhaps most unusually, Dow told the production managers and technicians who actually produced the plastic packaging, pesticides, drugs and myriad other products at the complex's 40 plants that they would have to present their pollution prevention projects directly to the environmentalists for approval and, once the projects had been started, to show them their progress reports.

As months went by, the apprehensive Dow managers were surprised by the cordiality of the environmentalists and their interest in the intricacies of the businesses. The environmentalists were not always equally relieved, especially when listening to managers who they said seemed uninterested in the impact of toxic chemicals as long as their operations were complying with all environmental regulations. For all the tension, though, the exercise bore fruit. Dow clearly wanted the Michigan Source Reduction Initiative, as the project was formally known, to succeed. It took pains to get that message across to the mid-level managers involved.

Dow is laying plans with Mr Bilkovitch and Ms Greer to duplicate the Midland experiment at its even larger petrochemical complex in Freeport, Texas. One hope is that, next time around, Dow and the environmentalists can achieve a goal they call 'institutionalisation'. Basically, that means measurably changing the corporate culture so that low-level managers routinely look for the kinds of improvement made here.

ENVIRONMENTAL ACTIVISM

Non-governmental organisations teaming up with business for a specific goal

Collaboration between environmentalists and business is controversial: such alliances can be seen as opportunistic on both sides, or as selling-out on the part of environmentalists. Collaborations are, however, increasingly common, although the temporary partnership in Germany between Greenpeace and Foron Household Appliances may be unique. The authors suggest that each such alliance is unique, although there are always lessons to be learned for application elsewhere.

Dr Cathy L. Hartman and Dr Edwin R. Stafford are both Associate Professors of Marketing in the Department of Business Administration at Utah State University in Logan, UT, USA. Dr Michael Jay Polonsky is Senior Lecturer in Marketing, University of Newcastle, Australia. This article appeared in a longer version in *Greener Marketing: A Global Perspective on Greening Marketing Practice,* which was published by Greenleaf Publishing in 1999. The authors are all academics, and the article reflects the academic point of view, use of language and style of writing. The story it tells is fascinating.

GREEN ALLIANCES
Environmental groups as strategic bridges to other stakeholders

Cathy L. Hartman, Edwin R. Stafford and Michael Jay Polonsky

The nature of business–environmentalist relations is changing (Lober 1997). Traditionally, environmentalists believed that the most effective means of enforcing corporate environmental responsibility was to adopt a singular, adversarial posture toward business (e.g. Murphy and Bendell 1997). In turn, firms viewed environmentalists as important stakeholders that needed to be considered, but kept at arm's length. While many firms still see environmental groups as a potential strategic threat, environmental groups can aid green marketing initiatives through various types of 'green alliance': collaborative business–environmental group partnerships that pursue mutually beneficial ecological goals (Stafford and Hartman 1996). In green alliances, environmentalists assist marketers *directly*, by providing expertise and technology, and *indirectly,* by influencing and brokering corporate relationships with other stakeholders to support the firm's overall green marketing programmes (Polonsky 1996).

Understanding, encouraging and managing these indirect relationships created by environmental groups can build 'strategic bridges' between the marketer and other stakeholders for corporate benefit. Strategic bridging refers to situations in which a third party links diverse constituencies together to address some broad-based problems that require input from multiple stakeholders, such as corporate environmentalism (Westley and Vredenburg 1991). Consider the strategic bridges created by Greenpeace between its corporate partner and other societal stakeholders in its campaign against the use of chlorofluorocarbons (CFCs) in Germany.

In 1992, Greenpeace championed 'Greenfreeze', an environmentally friendly hydrocarbon refrigeration technology, as a substitute for Freon, a leading CFC damaging to the ozone (Beste 1994; Kalke 1994). The 1987 Montreal Protocol on

Substances that Deplete the Ozone Layer mandated elimination of most forms of CFCs by the end of the 1990s, and scientists from Greenpeace and the Hygiene Institute, Dortmund, collaborated to develop the hydrocarbon technology as a viable alternative (Beste 1994). Although this hydrocarbon technology had been around since the 1930s, appliance makers had not considered it, due to its potential flammability. While modern refrigeration advances had eliminated this risk, major German appliance manufacturers were not interested in an old-fashioned, widely available technology that could not be patented to create a competitive advantage (Vidal 1992). Only the former East German manufacturer DKK Scharfenstein, later renamed Foron Household Appliances, was willing to consider experimenting with the hydrocarbon technology. However, unification had left Foron on the verge of bankruptcy. Foron, like many former East German firms, had been left with older, outmoded and less efficient technology which left them under-prepared to compete in a Western free-market system. As a result, the plant came under the control of the German privatisation agency, Treuhand. However, Foron's financial condition was such that, if investors could not be secured, the firm would be dissolved. With its engineers eager to save their jobs, Foron agreed to work with Greenpeace as a last resort to save its manufacturing operation.

After extensive talks in July 1992, Greenpeace granted Foron DM 27,000 ($17,000) to produce ten prototype hydrocarbon refrigerators. Racing against a liquidation timetable, Greenpeace and Foron fought a war of nerves with Treuhand who tried to block the project. At a stormy press conference showdown, Treuhand reluctantly allowed the Foron–Greenpeace project to proceed. The successful prototype, branded the 'Clean Cooler', not only won over Treuhand, but resulted in them providing Foron with substantial financial assistance and support in securing private investors. In March 1993, Foron's 'Clean Cooler', using Greenfreeze CFC-free technology, made its market debut (Walsh 1995).

To assist in developing the market for this product Greenpeace pre-sold the refrigerator to German consumers. This alarmed Western German refrigerator makers, who feared a substantial loss of market share. In an attempt to slow preproduction orders, these competitors launched a disinformation campaign through the trade press, warning retailers that Foron's Clean Cooler was 'an unacceptable danger in the home' and 'a potential bomb in the kitchen' and that Greenfreeze was 'energy-inefficient' (Vidal 1992). Greenpeace was charged with being irresponsible and obstructing constructive efforts to find feasible environmental solutions (*Air Conditioning, Heating and Refrigeration News* 1993).

Greenpeace's grass-roots publicity and product endorsement, however, generated over 100,000 orders in less than a year in Germany alone (Vidal 1992). One by one, the negative charges were reduced or dropped as Greenpeace's advocacy motivated the scientific community to align with Foron and the Clean Cooler against the chemical lobby. Later, Foron's Clean Cooler won the German Environment Ministry's prestigious 'Blue Angel' award. By 1994, all German refrigerator manufacturers had either switched to Greenfreeze CFC-free technology or were planning to convert, fulfilling Greenpeace's environmental goal of eliminating CFCs in German refrigerators.

Greenpeace directly assisted Foron, by providing both technical and financial support for its activities. While the direct aid for Foron is noteworthy, it was Greenpeace's indirect assistance, or strategic bridging, that proved indispensable to Foron. For the vulnerable Foron, Greenpeace brokered and negotiated a **political bridge** by influencing the Treuhand privatisation agency and a **credibility bridge** by influencing customers and the scientific community to support the Clean Cooler. Without Greenpeace's strategic bridging efforts, Foron would not have acquired critical stakeholder support and would have most likely gone bankrupt. This case demonstrates that environmentalist partners can become valuable bridging agents, providing firms with the necessary sociopolitical linkages and credibility for green marketing programmes that appeal to consumers and society at large (Mendleson and Polonsky 1995; Stafford and Hartman 1996).

Environmental groups act as bridging agents by forwarding their own ends *while at the same time* serving as links between the marketer and other stakeholders (Westley and Vredenburg 1991). As such, strategic bridging enables the development of environmental solutions especially when diverse stakeholders are unable to negotiate or co-operate freely because of mistrust, tradition, logistic problems or when there is a need for a third party to restore a balance of power, resources and expertise (Sharma *et al.* 1994). In particular, green alliances facilitate opportunities for marketers to address or satisfy demands of multiple stakeholders simultaneously through the crafting of **enviropreneurial strategies**—entrepreneurial innovations that simultaneously integrate and satisfy environmental, economic and social objectives for corporate benefit (Menon and Menon 1997). Foron's Clean Cooler is an example of an enviropreneurial product that benefited the environment, the company and society.

Enviropreneurial marketing: a multiple-stakeholder view of green marketing

An expanded, multiple-stakeholder view of green marketing has emerged called 'enviropreneurial marketing strategy', defined as 'formulating and implementing entrepreneurial and environmentally beneficial marketing activities with the goal of creating revenue by providing exchanges that satisfy a firm's economic and social performance objectives' (Menon and Menon 1997: 54). While traditional 'green marketing'—the production, promotion and reclamation of environmentally sensitive products (cf. Boone and Kurtz 1998)—has been the primary marketing response to consumers' concerns about ecological issues, global pressures from other sectors of society have escalated in the 1990s and are now also motivating shifts in firms' environmental behaviour (Roberts 1996).

Poor corporate environmental practices, for example, commonly lead to activist demonstrations, consumer boycotts and embarrassing media coverage (McCloskey 1992). Social and environmental performance have a strong bearing on the desirability of a corporation's stock and its access to credit (Gallarotti 1995; Sarkin and Schelkin 1991). Some corporate polluters are encountering difficulties recruiting employees from an increasingly socially aware workforce (Fischer and

Schot 1993) and, frequently, it is a firm's employees who are 'blowing the whistle' on their employers' environmental violations (Nixon 1993). Growing 'stakeholder activism' has made environmentalism a critical strategic issue for marketers in the global economy (Dechant and Altman 1994; Porter and van der Linde 1995), and improving stakeholder relationships and management are necessary for effective organisational performance across markets.

Emerging theoretical and empirical evidence indicates that environmental initiatives can lead to cost savings, increased profits and competitive advantages (Porter and van der Linde 1995; Russo and Fouts 1997). Menon and Menon (1997) assert that enviropreneurial activities can meet stakeholder interests at three levels:

- **Strategic level.** Strategic enviropreneurial initiatives reflect social responsibility and a desire to bring marketing activities into line with the expectations of current and future stakeholders. This may involve re-engineering corporate-wide activities for environmental sustainability. Marketers attempt to create a long-term, entrepreneurial prerogative through the development of new technologies, markets and products that create change within the industry or market.
- **Quasi-strategic level.** Quasi-strategic enviropreneurial initiatives are manifest in a desire to make corporate behaviour compatible with prevailing norms and expectations of critical stakeholders. Activities are modified in response to some immediate environmental need, and marketers attempt to integrate environmental concerns with strategy goals to achieve advantages within current businesses and markets.
- **Tactical level.** Tactical enviropreneurial programmes reflect social obligations to meet the minimum requirement to satisfy immediate market forces or legal constraints. Marketing mix variables are manipulated to position marketers/products as 'environmentally responsible', but they may provide limited environmental benefit.

While the specific approach (strategic, quasi-strategic and tactical) varies according to marketer, enviropreneurship is implemented so that the marketer considers how it can maximise the ability to address stakeholder interests (Menon and Menon 1997). Green alliances, such as the collaborative Foron–Greenpeace partnership, are catalysts of enviropreneurship.

'Collaborative windows' and green alliances

Stakeholder collaboration is part of a new trend in environmental problem-solving that has unfolded in the 1990s (e.g. Long and Arnold 1995). 'Partnerships' between nations, government agencies and the public and private sectors are becoming commonplace (Milne *et al.* 1996). This is due to the realisation that environmental problems are complex, transcending governmental boundaries. Moreover, the disparity of power and expertise among international stakeholders requires collaboration in order to deal with problems effectively.

Global concerns, such as ozone depletion and climate change, have encouraged co-ordinated international plans of action from political bodies (cf. National Association of Attorneys-General 1990), international trade organisations (e.g. the ISO 14000 series) and trade treaties (Levy 1997). Voluntary government programmes (e.g. the US Environmental Protection Agency's 33/50 programme) are stimulating over-compliance with environmental standards (Arora and Carson 1995) and flexible regulations (e.g. the 1990 US Clean Air Act's acid rain tradable credits programme) are encouraging firms to ally in order to develop and experiment with innovative enviropreneurial initiatives (Reitman 1997; Stewart 1993). The traditional 'command-and-control' character of environmental laws and regulations is shifting to market-based incentives and encouraging multiple-stakeholder collaborations (Hartman and Stafford 1997; Hemphill 1996). Further, private-sector social responsibility investment forums (e.g. Coalition of Environmentally Responsibility Economies [CERES]) are raising marketer accountability through environmental performance disclosure and institutional investor–environmental group collaboration (Wasik 1996).

Green alliances between marketers and environmentalists are perhaps the most unconventional outcomes of the 'collaborative window' because they involve formal co-operation between traditional adversaries (Hartman and Stafford 1997). Environmental groups commonly have scientific, legal and environmental expertise as well as public support for their activities (Lober 1997), and, as strategic partners, environmental groups can enhance the legitimacy of a marketer's enviropreneurial strategies. Greenpeace, for example, championed Foron's CFC-free Clean Coolers and used their network of support among retailers and final consumers to build pre-production market demand for the refrigerators. Green alliances also present opportunities for 'self-regulation' (Hemphill 1996), and environmentalists can facilitate negotiations with regulators and government agencies concerning corporate environmental activities, as demonstrated by Greenpeace's appeals to Treuhand on behalf of Foron. Thus, marketers are recognising that greater benefits can be achieved by working with environmentalists, rather than adopting a confrontational approach. Green alliances may bring about 'first-mover' benefits, both in markets and in addressing public policy concerns (Murphy and Bendell 1997). Likewise, environmental groups are increasingly turning away from traditional anti-business tactics for addressing environmental problems (Dowie 1995; Lober 1997; Mendleson and Polonsky 1995), although this shift in emphasis is not always embraced wholeheartedly on both sides.

The Foron–Greenpeace alliance marked the first time Greenpeace had backed a commercial product/technology and experimented with market-based principles to promote its agenda (Kalke 1994). Shortly after Greenfreeze's success in Germany, Greenpeace International's director Paul Gilding described the group's evolving environmental strategy, declaring, 'There's certainly a stronger predilection now for interfering in markets' (Levene 1994). In another statement, Greenpeace spokesperson Richard Titchen explained:

> We won't stop the actions that get much attention in the press and that have made Greenpeace famous, but now that people and companies

> have become more conscious of environmental problems, we consider it more effective to demonstrate solutions that are actually viable to industry (*Business and the Environment* 1994).

Greenpeace has proclaimed it will 'create new alliances with sectors such as businesses and industries' (*Business and the Environment* 1994), advocating technological solutions to environmental problems (Corder 1997). Using the market system, corporate collaboration, green technology development and stakeholder relations, Greenpeace and many other major environmental groups are recognising that the success of their activities is dependent on how well they satisfy or accommodate the needs of various stakeholders in society, including businesses (Hartman and Stafford 1997; Lober 1997; Mendleson and Polonsky 1995). Green alliances allow firms to become more involved in environmental solutions. While some criticise the use of green alliances as a 'sell-out' to business interests (e.g. Dowie 1995), Jay Dee Hair of the National Wildlife Federation has framed his organisation's industry collaborations by saying, 'We're not selling out, we're buying in!' (Dowie 1995: 75).

Strategic bridging and the stakeholder management process

In green alliances, an environmental group works on behalf of its corporate partner, leveraging its environmental credibility and advocacy image to bridge other society stakeholders and advance its corporate partner's enviropreneurial activities along with its own ecological agenda. Bridging organisations, while diverse, hold a common vision toward solving problems, making them particularly effective in contexts characterised by high interdependence and turbulence (Brown 1991). Additionally, because bridging organisations maintain their independence, they can negotiate bilaterally with relevant stakeholders. This freedom allows bridging participants the flexibility and opportunity to develop interpersonal familiarity that may eventually break down social and institutional barriers that typically separate stakeholders.

Through their strategic bridging capabilities, environmental groups can wield extraordinary influence among sociopolitical and economic constituents. Green alliances offer a form of marketer collaboration that is uniquely suited to the development of enviropreneurial strategies that take into consideration diverse stakeholders. However, similar to other types of alliance, it is essential that firms find the most appropriate environmental partner. Forming alliances with 'radical' environmental groups might fail to bring about the desired outcomes (Mendleson and Polonsky 1995). As in the case of the Foron–Greenpeace situation, the network of relationships between marketers and their stakeholders *and* between various stakeholders may be extremely complex.

Foron had a broad range of stakeholders who had direct and indirect effects on marketing activities. Key stakeholders who directly affected Foron included the Treuhand privatisation agency and private investors, who controlled financial resources; Greenpeace, who held scientific expertise and some financial resources;

the scientific community, who held technical knowledge; and consumers, who wanted environmentally responsible refrigerators. There were also a number of important indirect stakeholders, including policy-makers and supporters of the Montreal Protocol, who gave Greenpeace the impetus to collaborate with Foron; competitors, who undertook a media campaign of disinformation and scare tactics to negatively influence appliance dealers and consumers; and the media, who actively promoted both sides of the debate from information provided by various sources.

In the Foron–Greenpeace alliance, the government had a high stake because, without its state-funded support, even grudgingly given, the firm would have ceased to exist. Other stakeholders had significant indirect effects. Competitors, through the media, tried to scare off potential distributors and consumers, claiming that the new technology was dangerous. However, it should be noted that stakeholders' interests are dynamic. In the Foron–Greenpeace case, the media went from generally opposing Greenfreeze to supporting it. This shift only arose because of various indirect influencing forces, many orchestrated by Greenpeace. Thus, changes in stakeholders' behaviours or perceptions require monitoring and managing by the marketer and/or its bridging partner.

For the Foron–Greenpeace alliance, Greenpeace's actions to encourage Treuhand to extend state support to Foron conflicted with competitors' interests, and the threat of Foron's Clean Cooler sparked a disinformation media campaign in retaliation. Focusing on any one stakeholder may alienate another. Thus, enviropreneurship requires that marketers and bridgers understand and address complex stakeholder interests.

In strategic bridging, partnering environmental groups may employ (various) strategies to engage relevant stakeholders on behalf of their marketer partners. Some of the stakeholder strategies employed by Greenpeace to support Foron are discussed in turn.

For example, Greenpeace used an **adaptive** strategy in working with Foron. The adaptive approach is where a strategic bridging agent attempts to modify its own behaviour or that of its marketer ally to meet another stakeholder's interests (Polonsky 1996). The financially distressed Foron adapted its product to meet both Greenpeace's technological recommendations and the Montreal Protocol standard because it perceived this technology as a way of ensuring its viability. At the time, Foron was no more environmentally committed than other refrigerator manufacturers, but Foron engineers perceived the CFC-free Greenfreeze technology as a potential market opportunity. Eventually, after the Clean Cooler's success, Foron committed itself to a long-term enviropreneurial market position. In late 1994, Eberhard Gunther of Foron announced, 'We want to stand our ground with intelligent, innovative and, above all, ecological appliances' (Kalke 1994: 24). Energy efficiency became Foron's enviropreneurial focus and, later, another innovative refrigerator design was developed which further reduced the appliance's energy consumption, thus making it even less environmentally harmful (Kalke 1994).

Greenpeace also used an **aggressive** strategy among some stakeholders. This is where a strategic bridging agent attempts to change the other stakeholder's views directly or the other stakeholder's ability to influence the firm's marketing outcomes (Polonsky 1996). Examples of activities that could be used in an aggressive strategy include: public advocacy, educating target groups, or organising protests. Each of these may have varying levels of effectiveness in changing a stakeholder's beliefs and/or behaviours and will depend on the stakeholder's interests in relation to the specific situation under consideration.

When Treuhand attempted to block the production of the ten prototype hydrocarbon refrigerators, Greenpeace orchestrated a press conference showdown. This aggressive action forced Treuhand to allow Foron the opportunity to continue its operation. Greenpeace's aggressive action directly challenged the authority of the privatisation agency. While Greenpeace's aggressive approach was successful in making Treuhand rescind its dissolution decision, it may *not* have facilitated a change in Treuhand's mind-set. That is, it is unclear if Treuhand believed the Foron–Greenpeace partnership would result in desirable outcomes for the firm or environment. The use of an aggressive approach to change stakeholders' beliefs may be more successful when bridging groups assume a formal public advocacy role rather than an overtly antagonistic one. Changing a stakeholder's beliefs may take more time than changing an immediate behaviour.

When a bridging group decides a stakeholder belief change is warranted, it may use an aggressive approach in an indirect manner (i.e. by influencing other stakeholders to exert pressure) rather than a direct manner. For example, prior to the Foron–Greenpeace alliance, Greenpeace appealed directly to a number of German appliance manufacturers to consider Greenfreeze. After they refused, Greenpeace allied with Foron, demonstrating that Greenfreeze could be developed into a viable product. Greenpeace then attempted to change the industry's perceptions about Greenfreeze by appealing to consumers and the media, who, in turn, engaged the scientific community to support the technology. These actions indirectly influenced the industry eventually to accept Greenfreeze. Thus, Greenpeace's aggressive strategy was enacted indirectly, stimulating consumer demand and changing media disclosure, which, in turn, pressured industry to modify its beliefs and behaviour.

Before the Foron–Greenpeace alliance, Greenpeace had employed a **co-operative** strategy with scientists at the Hygiene Institute to develop the CFC-free Greenfreeze technology. A co-operative approach is where the strategic bridging agent works with a stakeholder to achieve a desired set of outcomes (in this case, an alternative refrigeration technology to meet the Montreal Protocol). Co-operation ensures that the interactive nature of the marketer and its stakeholders is recognised (Polonsky *et al.* 1997; Rowley 1997), and solutions can be designed as long-term strategic adjustments. Greenpeace's Greenfreeze campaign co-ordinator, Wolfgang Lohbeck, had worked as a department head in Lower Saxony's Ministry of Science before joining Greenpeace (Beste 1994). His personal connections within the scientific community ultimately helped to leverage its support for Greenfreeze, confirming its environmental superiority and safety. This bridge, in turn, encouraged consumer demand and government endorsement for Foron's Clean Cooler.

Risks of green alliances and strategic bridging

Despite advantages, environmental groups engaged in green alliances as strategic bridging agents present many challenges both to themselves and to their stakeholders. Groups serving as strategic bridging agents between the firm and its other stakeholders need 'to obtain "back-home" commitment from their constituents—because it remains at all times an independent entity with its own agenda' (Sharma *et al.* 1994: 461). That is, the bridging group's stakeholders must support the bridging group's activities. In this way, environmental groups must find mechanisms to integrate other stakeholders who may possess divergent values, interests, wealth, power, culture, language and structural characteristics (Brown 1991). The more divergent stakeholders' interests are, the more difficult the bridging problem may be. Further, members of the environmental group must understand the diverse stakeholder perspectives that its leaders are trying to integrate; any internal group contentions that arise from enviropreneurial programmes that represent 'compromised' solutions to broader environmental problems will seriously weaken an environmental group's strategic bridging ability (Westley and Vredenburg 1991). For a bridging agent to be successful, it must effectively market its partner's cause and become an advocate for the specific enviropreneurial marketing activity to its members, without tarnishing its own reputation.

This bridging challenge is highlighted in a Canadian green alliance, initiated in 1989 between Pollution Probe and grocery retailer Loblaws. The environmental partner attempted to link environmental stakeholders and consumers to the retailer via a green product endorsement. Many of Pollution Probe's staff and membership disagreed with the endorsement, and Greenpeace questioned publicly the 'environmental soundness' of the endorsed products. The media frenzy that followed damaged Pollution Probe's credibility and its ability to bridge necessary stakeholders (see Stafford and Hartman 1996; Westley and Vredenburg 1991).

Marketers also face risks. Collaboration with environmental groups can offend core customer segments if they view environmentalists negatively as extremists (cf. Rohrschneider 1991). Moreover, green alliances can propel marketers and their enviropreneurial programmes into the public spotlight, encouraging scrutiny from other critical or threatened stakeholders, as exemplified by Foron's competitors' reaction to Greenfreeze's technological challenge. Central to the success of strategic bridging is the compatibility of co-operative objectives and vision between the marketer and the environmental group (cf. Brown 1991). Over time, as partners' objectives are met, or as they change or diverge, the strategic bridging partner may become less willing to broker and negotiate linkages between the marketer and other domain stakeholders, potentially jeopardising the marketer's competitive advantage (Westley and Vredenburg 1991). Consider the events following the successful product launch of Foron's Clean Cooler.

By 1994, Foron and Greenpeace were not surprised when West German appliance manufacturers switched to the hydrocarbon technology (Kalke 1994). For Greenpeace, the industry's adoption of its Greenfreeze technology was the realisation of its primary campaign objective of eliminating CFCs in German refrigerators. With the marketing experience gained from the Foron alliance, Greenpeace introduced the hydrocarbon technology to China, India and other developing countries (Beste 1994). Greenpeace literally gave the technology to willing enviropreneurs, convinced that Greenfreeze, if readily available, would be adopted widely in the developing world.

For Foron, however, the German industry's adoption of Greenfreeze presented a grimmer marketing reality. Being the first to develop the technology commercially did not lead to Foron developing a long-term competitive advantage, as the technology was not exclusively held by them. This resulted in the other, better-resourced firms marketing their versions of environmentally responsible refrigerators more effectively. Thus, despite its initial success, Foron's line of Clean Coolers did not rescue the firm from its financial crisis, and Foron's market share eroded as more sophisticated, rival hydrocarbon refrigerators appeared on the market. In 1995, Samsung entered into negotiations to buy the company, but bowed out after six months, citing that Foron did not fit with its planned European strategy (*Handelsblatt* 1995). Shortly after, Koc of Turkey began acquisition negotiations, only to withdraw the following year due to Foron's poor sales and financial situation (*Handelsblatt* 1996). Greenpeace had already abandoned the company to concentrate on its global Greenfreeze campaign, and Foron lacked the financial resources and marketing know-how to establish itself independently. In March 1996, Foron declared bankruptcy (*Die Welt* 1996).

The aftermath of the Foron–Greenpeace alliance illustrates a key risk associated with strategic bridging. Because strategic bridgers are motivated to collaborate and engage other stakeholders on behalf of their corporate partners through forwarding their own agenda (Westley and Vredenburg 1991), once that agenda is met or is no longer being served through collaboration, the bridger will no longer be committed to its corporate partner. Understanding the nature of strategic bridging through green alliances warrants consideration of partners' values and goals (cf. Rokeach 1973). Fundamentally, environmentalists and businesses hold diverging, if not conflicting, goals and/or even values. For environmentalists, ecological goals are foremost, whereas, for businesses, profit and market objectives are paramount. In general, green alliances represent lower-level instrumental (or means) values for their partners—potentially desirable mechanisms for achieving each individual partner's goals (cf. Lober 1997). Thus, there is frequently little long-term commitment by either partner to the other or the other's agenda. In these situations, the environmentalist partner is likely to exit the relationship when the alliance no longer meets its ecological goals. If this breakdown occurs before the corporate partner has achieved its objectives, linkages with critical stakeholders may be placed at risk, as the firm no longer has an effective bridge with them.

Via the Foron–Greenpeace alliance, Greenpeace had introduced an alternative refrigerator technology throughout the German refrigerator industry, which had been opposed by the industry. Therefore, it no longer needed to continue supporting Foron, even though Foron had not resolved its financial dilemma. Though speculative, it is conceivable that Greenpeace *would* have assisted Foron in bridging and procuring investors if establishing consumer acceptance for Clean Coolers had taken longer *or* had the industry delayed its adoption of Greenfreeze; either scenario would have required Greenpeace to continue helping its cash-strapped partner by bridging necessary stakeholders, including investors, for product success. Wolfgang Lohbeck of Greenpeace observed, 'It was a piece of luck that we could win one company over to our way of thinking and that this firm could turn facts quickly into marketable realities' (Beste 1994: 29). Perhaps Foron's products were too successful in that their immediate market demand signalled Greenfreeze's market opportunities to competitors and constrained the time Foron needed to leverage Greenpeace's bridging capabilities to investors to remain competitive. After the industry-wide conversion to Greenfreeze, assisting Foron was no longer instrumental to Greenpeace's agenda.

Social concerns with green alliances

Beyond the managerial implications involved in green alliances, a number of broader social issues warrant consideration. Foremost is the need to ensure that environmental groups are not perceived as 'caving in' to business interests (e.g. Dowie 1995) and thus seen no longer to be objective in relation to given environmental issues. Environmental group initiatives with business partners are typically expensive and time-consuming, and the results for the environment usually represent compromised, incremental progress rather than significant changes in corporate environmental behaviour (Murphy and Bendell 1997). Adversarial protests and tactics that rally stakeholders against environmentally detrimental corporate practices can frequently instigate sweeping changes, as exemplified by the Dolphin Coalition's consumer boycott of tuna that catalysed industry-wide 'dolphin-safe' fishing practices (Stafford and Hartman 1996). Are the traditional campaigns of protest becoming obscured by closer business ties? Environmentalist enthusiasm for green alliances may be sidelining other priorities aimed at broader society, such as consumption reduction, product re-use and recycling.

Another concern centres on whether a partnering environmental group should be playing the role of a 'free' environmental consultant. As noted earlier, the incentive of an enviropreneurial green alliance strategy for a marketer is either to improve internal efficiencies or to develop differential advantages over competitors (cf. Lober 1997; Menon and Menon 1997; Porter and van der Linde 1995). Is it appropriate for a non-profit environmental group, who may rely on grass-roots donations from the general public (consumers), to support just *one* marketer in a green alliance? Are the broader environmental/societal benefits resulting from green alliances outweighing the potential marketer advantages (e.g. price premiums) that may be gained over the general public?

Lastly, there is the issue of transparency (Murphy and Bendell 1997). The co-operative relationship between a marketer and its environmentalist partner requires a degree of confidentiality. Many firms have misgivings about opening their operations and marketing processes to environmentalists for fear that information may be used against them if the green alliance fails. Agreements on confidentiality are necessary to elicit marketer participation, but they may conflict with the environmental group's accountability to its membership and its 'watchdog' role in society. With increased emphasis on green alliances, 'there is a danger that the environmental group may begin to adopt the kind of paternalism normally associated with "establishment" institutions' (Murphy and Bendell 1997: 129).

ENVIRONMENTAL ACTIVISM
Bottom-up change

Employee involvement in changing the environmental performance of businesses is stimulating new thinking, with the potential for new power relationships. Following is a contribution by Ole Busck, who is Danish, written as a result of his experiences as an official in Denmark's largest trade union. He describes what is happening in Europe in employee involvement as well as the sometimes unexpected potential of the new dynamic.

Originally a shipbuilder at Elsinore shipyard in Denmark, Ole Busck took a MA in political and social science and for 15 years was employed in SiD, the General Workers' Union of Denmark, as health, safety and environment expert. In the course of the growing importance of environmental development in enterprises, he played a central role in the building of understanding between trade unions, environmental authorities and progressive companies. At present, Busck works in a private environment consultancy, PlanMiljø, advising private and public enterprises on clean production and environmental management, including employee participation.

EMPLOYEE PARTICIPATION
An important resource in environmental development

Ole Busck

Scientific studies and reports from experienced companies are confirming what progressive managers and trade union officials have alleged for years: when environmental development takes place in companies, major benefits accompany the involvement of employees in the process of change.

Environmental management as an instrument of change in companies requires responsibility, understanding and skills from employees, and such changes in turn benefit from employees' influence on decision-making. Businesses seeking environmental improvements have experienced essential results by virtue of active employee participation in innovations at all levels, from changing simple work routines to the elaboration and implementation of environmental policy throughout the company. Case studies in companies and industries have documented that employees possess a fund of knowledge that can be activated and integrated into the environmentally driven transformation of work organisation, procedures, management processes, process and product design (Handberg 1993; Busck and Handberg 1995; LO 1998; Lorentzen *et al.* 1997; ÖTV-Union 1995; Christensen *et al.* 1996; Zwetsloot and Bos 1998; Cerne and Antonsson 1999; Dalton 1998; Kamp 1997; Gee 1994; Kofoed *et al.* 1998).

It is a basic finding that the motivation and interest of employees must be aroused in order to carry out necessary changes in work routines. This implies working with **attitudes**. The more open dialogue there is between management and employees on objectives and procedures, the more likely it is that environmental initiatives will be embedded in the organisation and in actual performance at the workplace. Experiences and attitudes from everyday private and public life of employees may be exploited in the greening of an enterprise. Another precondition to achieve co-operation with thorough change in a company is **participation**. The real power of co-decision-making can release the creativeness and involvement of employees.

Studies on the introduction of environmental management systems in companies in European countries have shown that involvement of employees is a key factor in determining whether the new system remains at the level of bureaucratic paper routine (and maybe was only intended as such) or becomes instead anchored in the organisation, thereby leading to permanent changes in environmental performance. In the end, the concept and use of EMS (environmental management system) as a management tool is decisive.

There is no doubt that from the point of view of employees in general, there are few obstacles to being involved in decision-making regarding environmental matters. A recent Danish survey among employees (a representative cross-section of members of LO, the national trade union federation) showed that 50% were already participating in the environmental work of their company and that more than a third of those remaining wished to be more involved.

It is important to realise, however, that to the extent employee resources can be activated and can contribute to environmental improvements, the demand for improvements in the **working environment** is also activated. Whether management has planned to integrate health and safety concerns into the environmental initiative or not, the actual involvement of employees means that health and safety issues are put on the agenda.

Research has provided many examples of management-focused environmental initiatives in companies with no effects on the working environment (often not leading to qualitative improvements in environmental performance either), but few examples of environmental initiatives involving employees that do not encompass health and safety and lead to simultaneous improvements in the working environment.

What is particularly interesting is that a **synergy effect** is often observed when employees are involved in the environmental development of their company. Not only does the fact that work-environment problems are included in the process lead to enthusiasm on the part of employees in the company's environmental work, in addition, a creative dynamic is established in the health and safety work of the company.

Work with environmental improvements brings about solutions in common problem areas such as spills and accidents, hazardous chemicals and so on; improvements can be designed to meet many concerns, thus creating synergy effects. Further, the engagement of employees in the process—their experience of being attended to, their sense of the possibility of real change—can eliminate traditional barriers to change in the work environment. In the case of repetitive work, where the psycho-social load connected with fixed-station work and low influence is heavy indeed, involvement in processes of environmental change has shown a potential for breaking down walls.

Government-funded Scandinavian research has documented the results in environmental performance as well as improvements in work environment from actively involving and using the creativeness of employees. In the Danish fish-processing industry, characterised by physical routine work and heavy psycho-social burdens, it has been possible, through bottom-up innovations and worker participation in the development of technology, to reduce injuries significantly. At the same time, remarkable improvements in emissions of waste and use of resources have been achieved, along with improved company earnings.

It is a central point in the research that the existence of a **democratic situation** at the workplace is seen as a precondition for activating employee resourcefulness. It is not enough to inform and put up a suggestion box. It is insufficient to allow participation only for a few employee representatives. Involvement in a wide sense is necessary to facilitate the necessary shift in attitudes and engagement.

As another important precondition, the studies (similar to a series of surveys published by unions in several countries) underline the existence of possibilities for **qualification- and competence-building**. Education is an essential means of action, not only to improve qualifications but also to prepare people for new attitudes and roles.

Barriers to participation

Management must accept the apparent transfer of power embodied in the precondition of a democratic forum. The democratic values of society must be respected on the factory floor. All parties have to engage in the building-up of a learning organisation, with respect for employees, their abilities and their personal development.

Traditional attitudes of employees are also a part of the industrial culture and are sometimes an obstacle. Attitudes such as: 'We are only in the job to earn money' or 'All responsibility for my work environment and the environment of the company is in the hands of the employer' are obstructive and irrelevant. Surveys among union members in European countries have made it clear that attitudes of this kind are diminishing.

Means of action

The evolution of employee resourcefulness and participation is first and foremost a process of development within a company. But it is important that authorities and relevant institutions of society contribute to this development and work to improve the preconditions for it and support it:

- **Legal initiatives.** Two recent initiatives, one Dutch and one from the European Union are giving important signals to companies about involving employees:
  - In the Netherlands, it has been decided that company committees must be consulted regarding questions of environmental performance of the enterprise.
  - In Denmark, the 1997 Act on Green Accounting requires employee participation when safety and health issues are included. In connection with the recent implementation of the EU IPCC (Intergovernmental Panel on Climate Change) Directive, a provision has been

added in the Danish Environment Protection Act, requiring that 'enterprises to be approved must organise environmental work in a way that ensures the involvement of employees'.

- **Agreements.** In some European countries, agreements exist regarding employee representation on environment committees in enterprises, but no examples of larger joint activities engaging employees in a wider sense, or of building up democratic and learning organisations connected with environmental, safety, and health issues have been reported.

  Some social partners in specific sectors, e.g. the Swedish paper and pulp industry and the German hotel and restaurant sector, have been working on the issue of 'green agreements' for some years. No documentation is available on how far they have come in changing traditional behaviour and to what effect.

  The power of large industrial organisations to inform and influence their members and to lobby environmental authorities is great. Such large organisations could certainly be used more effectively to promote employee participation, but, in general, they do not seem inclined to consider sustainable development and the role of employees in the process as core business.

  From Denmark, a single example of a constructive initiative has been reported. Based on the results from the project on worker involvement in the fish-processing industry, the metal industry in 1995 initiated the setting-up of a model educational system as part of the national vocational-training system to train employees in environmental action. The training incorporates participation in the environmental development within the specific business of the employee. The training is very popular among environmentally progressive enterprises in Denmark.

Some work has been done to develop methodology for activating employee resources, competence-building and involvement in environmental development. From Finland, the TUTTAVA methodology, developed by Professor Jorma Saari, aimed at worker participation in 'green housekeeping' in enterprises, is reported to be popular among Finnish enterprises.

Attention and support from political authorities to the further development and dissemination of these methods in enterprises would help to develop the dynamics in the industrial culture necessary to make environmental development a joint and integrated process of sustainable development.

INTERNET ACTIVISM

Assembling large numbers of environmental activists at the December 1999 WTO (World Trade Organisation) meetings was made possible largely by the Internet. Many say the strength of such protests will only grow with the spread of the World Wide Web. For instance, a national training camp for environmental protesters was organised in the US during the spring 2000 school break via the Internet.

As an example of the Internet's potential to stimulate public participation via information provision, the US non-governmental organisation Environmental Defense (formerly Environmental Defense Fund)'s website Scorecard[26] is an interactive platform based on right-to-know laws and shows toxic chemical releases from 21,000 manufacturing facilities in the US. The data is from the US Environmental Protection Agency Toxics Release Inventory (TRI), which includes 650 chemicals, heavy metals and toxic wastes, and includes information on the potential health hazards of toxic chemicals. The TRI system was created under the Emergency Planning and Community Right to Know Act of 1984, following the Union Carbide chemical release disaster in Bhopal, India.

Environmental Defense's Scorecard website allows you to use a pollution locator to find polluting companies and their locations. With Scorecard, you can access a map of each state showing toxic chemical releases.

Examples of pages from the Scorecard are shown in Figures 54–58.

In the UK, the non-governmental organisation Friends of the Earth has a website called Factory Watch,[27] which gives facts about industrial pollution in that country. As with Scorecard, you can obtain official pollution figures for local industries in England and Wales (you cannot get Factory Watch-type information on major industry in Scotland or Northern Ireland, because the data is not collected).

The US Environmental Protection Agency has a website called Envirofacts,[28] which allows users to view state-by-state TRI reports as well as carry out searches based on location, chemical names or facility names.

The UK Pollution Inventory data is accessible from the UK Environment Agency's website[29] which also features a Hall of Shame of companies that have been found guilty of environmental violations.

26 www.scorecard.org/about/about.tcl
27 www.foe.co.uk/campaigns/industry_and_pollution/factorywatch/
28 www.epa.gov/enviro
29 www.environment-agency.gov.uk

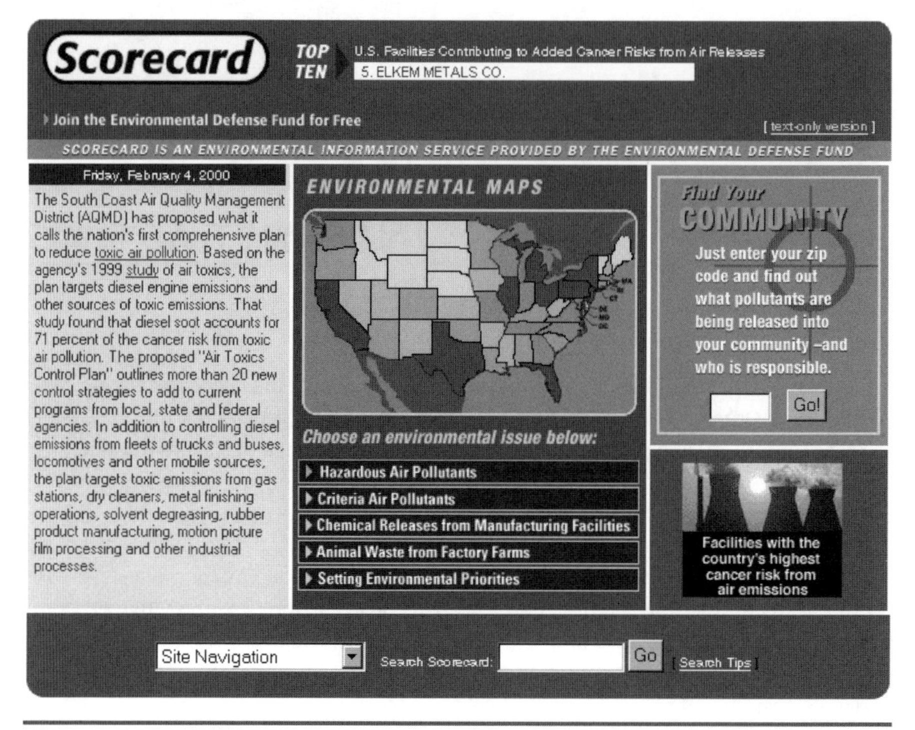

Figure 54 **Scorecard: main page**

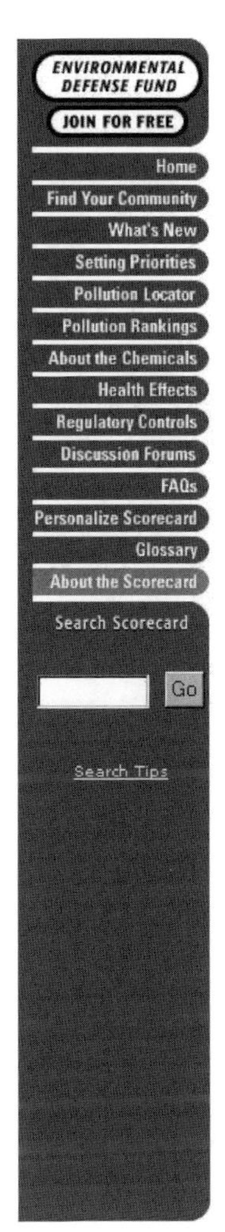

ABOUT SCORECARD | Main Page

Scorecard is the ultimate source for free and easily accessible local environmental information on the Internet. Simply type in a zip code to find out about local air pollution and explore state-of-the-art interactive maps. Scorecard delivers accurate information on the toxic chemicals released by manufacturing facilities and the health risks of air pollution. It can rank and compare the pollution situation in areas across the US. Scorecard also profiles 6,800 chemicals, making it easy to find out where they are used and how hazardous they are. Using authoritative scientific and government data, Scorecard provides the most up-to-date and extensive collection of environmental information available on the web today. Information is power - once you learn about an environmental problem, Scorecard encourages and enables you to take action - you can fax a polluting company, contact your elected representatives, or volunteer with environmental organizations working in your community.

Using the Scorecard

Learn How to Use Scorecard - Even if you don't know about the Internet, our Scorecard Guide will steer you easily to the information you want about the environmental health of your community. Learn how to get the real story on toxics in your community and use what you know to organize a community meeting on toxics and reduce environmental hazards in your area.

Sample Uses of Scorecard - How Scorecard can help answer several commonly asked questions regarding pollution and toxic chemicals.

Why Scorecard? - EDF's Executive Director, Fred Krupp, explains why EDF created Scorecard and how it can strengthen the right-to-know movement.

Caveats - Important facts you should know about Scorecard's content and what Scorecard can and cannot tell you.

Use of Scorecard Information - We'd like you to use Scorecard as a resource for information about toxic chemicals in the environment. Join the Scorecard Network of web sites that use our maps or data. Please review our instructions for copying and crediting Scorecard information.

Information About the Scorecard Project

Scorecard in the News - The latest press coverage of Scorecard.

Scorecard's Data Sources - Scorecard integrates over 300 scientific and governmental databases.

How Scorecard Works - The information technology behind Scorecard.

The People Who Created Scorecard - EDF staff and consultants who contributed to Scorecard's development.

Funding for Scorecard - Financial and technical contributors to Scorecard's development, and an invitation to support future Scorecard enhancements.

Future Improvements to Scorecard - New content, information technologies and interaction opportunities under consideration for future Scorecard versions; let us know what you would like to see.

Figure 55 **About Scorecard**

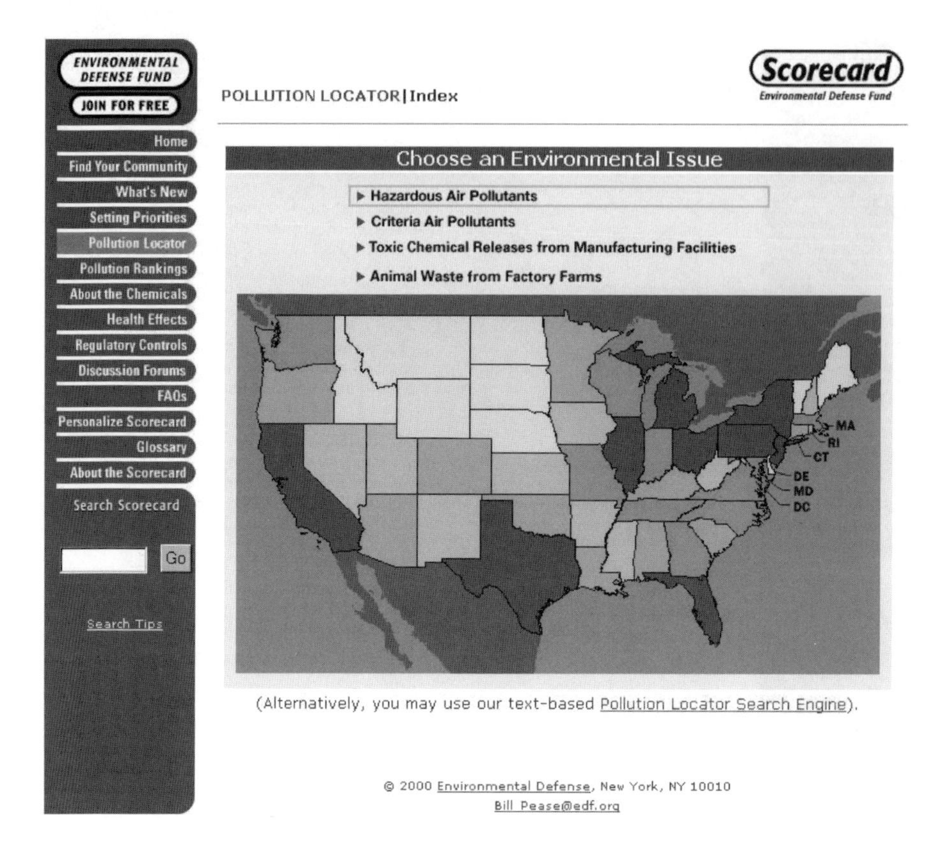

Figure 56 **Pollution locator: index**

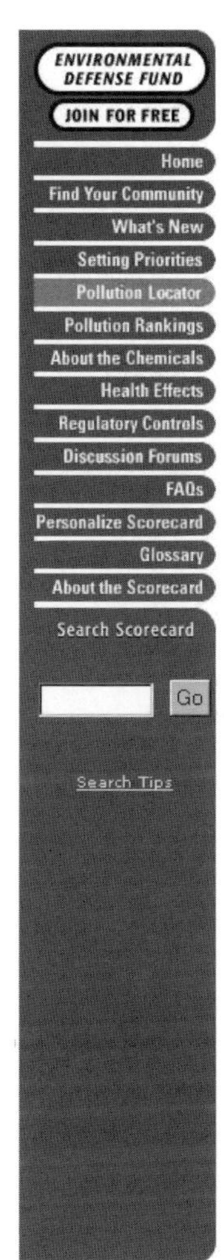

POLLUTION LOCATOR | Hazardous Air Pollutants

Hazardous air pollutants (HAPs) are chemicals which can cause adverse effects to human health or the environment. Almost 200 of these chemicals have been identifed, including chemicals that can cause cancer or birth defects. Very little is known about the potential health risks from this type of air pollution because fewer than 50 locations in the U.S. regularly measure the concentrations of HAPs in ambient air. To help identify air quality problems, the U.S. Environmental Protection Agency recently estimated the concentration of HAPs in every locality in the U.S. (over sixty thousand census tracts). Scorecard combines these EPA estimates with data on chemical toxicity to present a screening-level characterization of the cancer and noncancer risks posed by HAPs. See Scorecard's overview of hazardous air pollution problems.

United States
Number of People Living in Areas where the Estimated Cancer Risk from HAPs is Greater than 1 in 10,000

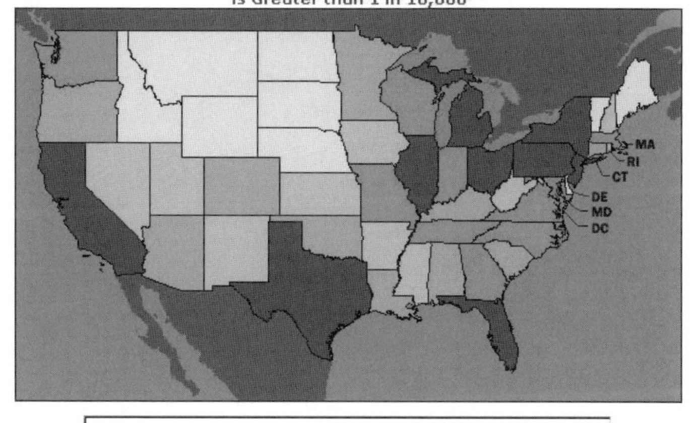

Map Legend:
Number of People Living in Areas where the Estimated Cancer Risk from HAPs is Greater than 1 in 10,000

- highest 20% of states
- second highest 20% of states
- middle 20% of states
- second lowest 20% of states
- lowest 20% of states

Hazardous Air Pollutant Reports Available at This Level

- View National Report

Choose a State | View Report

NOTE: Scorecard combines exposure data from the U.S. EPA Cumulative Exposure Project with health effects information to estimate the health risks posed by chemical pollutants in ambient air. EPA exposure estimates, and the Scorecard risk estimates that are based on them, provide a screening-level assessment of hazardous air pollution problems and are subject to important caveats. EPA's exposure estimates are based on 1990 emissions data, although they are generally consistent with current air monitoring data. Scorecard's risk estimates are calculations based on models: they are useful for ranking purposes, but are not necessarily predictive of any actual individual's risk of getting cancer or other diseases.

Figure 57 **Pollution locator: hazardous air pollutants**

- Map Locating Hazardous Air Pollution

CONNECTICUT

[top]

- Ranking Areas by Health Risk

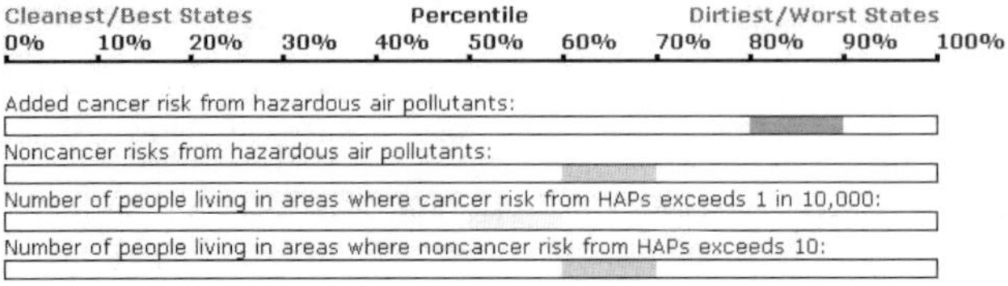

Cleanest/Best States	Percentile	Dirtiest/Worst States
0% 10% 20% 30%	40% 50% 60%	70% 80% 90% 100%

Added cancer risk from hazardous air pollutants:

Noncancer risks from hazardous air pollutants:

Number of people living in areas where cancer risk from HAPs exceeds 1 in 10,000:

Number of people living in areas where noncancer risk from HAPs exceeds 10:

Rank states or counties by health risk from HAPs.

***Note:** These rankings are based on exposure estimates derived from 1990 emissions data. There are important uncertainties associated with using models to predict current air quality. Scorecard risk estimates are based on EPA exposure estimates and provide a perspective on the magnitude and sources of hazardous air pollution problems. They are not definitive evaluations of health risk in a particular locale.

[top]

- Cancer Risks and Noncancer Hazards in CONNECTICUT

Cancer Risks from Hazardous Air Pollutants:

Average individual's added cancer risk:	310 per 1,000,000
Population in areas where cancer risk exceeds 10^{-3}:	55,809
Population in areas where cancer risk exceeds 10^{-4}:	3,267,303
HAP with the highest contribution to cancer risk:	POLYCYCLIC ORGANIC MATTER (POM)

Noncancer Hazards from Hazardous Air Pollutants:

Average individual's cumulative hazard index:	10
Population in areas where hazard index exceeds 1:	3,287,116
HAP with the highest contribution to noncancer hazards:	ACROLEIN

[top]

- Sources Contributing to Health Risks from Hazardous Air Pollutants

	Contribution to Added Cancer Risk	Contribution to Cumulative Hazard Index
Area source	26%	49%
Mobile source	40%	47%
Point source	34%	4%

See sources of specific Hazardous Air Pollutants

[top]

- What We Don't Know About HAPs in CONNECTICUT

17 hazardous air pollutants in CONNECTICUT lack the risk assessment values required for safety assessment

[top]

- Take Action

Send email to the U.S. Environmental Protection Agency
Send an email to decision makers in CONNECTICUT
Join an on-line discussion about CONNECTICUT
Volunteer with environmental groups in your community
Network with environmental groups

[top]

- Links

Scorecard reports on:
- Criteria air pollutants
- Manufacturing facilities (TRI)

[top]

Figure 58 **Map locating hazardous air pollution**

CAR SHARING

Car sharing is catching on. In many cities, from Seattle to Zurich, commuters or shoppers pick up small cars at handy stations for a modest fee, drive to the office or shopping mall, and leave the car at a nearby station for someone else to drive. After work or errands, the commuter or shopper picks up another car to go home, leaving the car at a station where it is available to go out for the evening with someone else at the wheel. Car sharing promises to become part of a multiple-transportation mode to get from here to there.

Michael Glotz-Richter is head of strategies for the environment in the Bremen City Department for Building and Environment as well as Co-ordinator of the Bremen ZEUS project, a European Union pilot programme for innovative transport. He is also a Green Party senator. In a country where the private automobile is a symbol of prosperity and identity and, where drivers speed along on autobahns at 200 kmh, it would seem impossible to make a dent in the automobile's ubiquitousness. Yet the Bremen car-sharing programme is catching on. A European Union study of car-sharing systems showed that, if all Europe were to have them, such programmes could reduce car mileage by as much as 32 billion km a year. This would reduce CO_2 emissions by as much as 5 million tons a year.

Michael Glotz-Richter offers the following story in which party politics, local, national and federated government, the public transport system, and a new enterprise find their full complementary expression.

CityCarClub/car sharing

Experience of a municipality with an innovative mobility service as a strategic move towards sustainable development

Michael Glotz-Richter

Our cities suffer from traffic not only in terms of noise and pollution—but also with the increasing use of road space by cars. While we have technical improvements related to emissions, the problem of cars dominating public space is increasingly horrific. Especially in the inner city mixed-use areas, the very limited road space leads to strong competition among all potential users.

Parking on the sidewalk is often common—any improvements for pedestrians (the sidewalks are also a playground for children) need to tackle the problem of parked vehicles. Every kind of greenery is in competition with parking, etc. To find political acceptance for improvements, intelligent solutions must be developed. Parking garages are not the solution—plus they are usually difficult to finance.

Development of car sharing in the European context

In the European Car-Free Cities Network, a working group, 'Practical Alternatives to the Car', was founded in autumn 1995. The working group succeeded in getting car sharing firmly on the agenda of the European Commission. A brochure, *CityCarClub/Car-Sharing: Carfree but Carefree*, printed in English, French and German, has been disseminated worldwide since its launch at the Car-Free Cities Copenhagen Conference.

The European Commission's Directorate for Transport's Green Paper, 'Citizens' Networks', which deals with public transport, envisioned an important role for car sharing as a support for the public transport system. Feedback from the Car-Free Cities network, contributions to the EU Citizen's Networks Forum, and official comments from the German government have all further supported that development. At an October 1999 conference on car sharing in Brussels, the location of the European Commission Headquarters, the first step towards launching an operation in Brussels was taken. Owing to the early arrival of car sharing in Bremen, the city has played an important role in the wider development of the concept across the EU.

Bremen: urban revitalisation needs an answer to the problems caused by increasing car ownership

Car sharing: Bremen

Bremen was one of the first cities in Germany to have a car-sharing system. Following successful examples in Switzerland and Berlin, StadtAuto Car-Sharing started in Bremen in 1990, first as a club with 28 participants sharing three cars. In the beginning, a high level of environmental commitment was central to the club's functioning. Participants almost felt they were making a sacrifice to help the environment. Public scepticism was rampant. Could such a concept work in a city such as Bremen, where DaimlerChrysler is the largest employer? The example of Switzerland was surely not transferable, said doubters: the Swiss are more reliable and care much more about sharing things than the Germans. With expansion and the development of a more market-oriented structure in the club, sceptics did not disappear, but they have less to say.

By the summer of 1999, Bremen's StadtAuto system had grown to almost 2,000 members, in a city of about 545,000 inhabitants. The fleet of StadtAuto cars had grown to more than 85 and, thanks to co-operation with an Opel dealer, more cars are available when necessary to meet peak demand, especially on weekends. The network of car-sharing locations where the cars are available has developed to more than 45, all over the city, and car sharing is now mentioned in city development plans, so it will be built into new developments.

As Bremen is a cyclist's city, with bicycles accounting for 22% of all trips by Bremen residents, car sharing has an important role in further supplementing environmentally friendly modes of transport, offering a real alternative to ownership of a private car. It is important that the daily routine (home-to-work trips, shopping, leisure trips) can be accomplished without a car, and car sharing permits this. It is in effect a mobility insurance policy for all cases when public transportation, walking and cycling are not adequate: at night, when public transport is not available or convenient, when bulky or heavy objects have to be transported, and so on. Surveys show that car sharers also use public transport more often.

Public transport season ticket plus car-sharing AutoCard

In addition to walking and biking, public transport can be a basic mode of transport. But a car gives flexibility, as it is not bound to timetables and routes. A combined system of public transport and car sharing may give the flexibility of car ownership without the drawbacks of expense and inefficient use of space.

A major project in Bremen is the *Bremer Karte plus AutoCard*, combining car sharing and the public transport services of the local operator, BSAG, to offer a full mobility service. The new intermodal service started on 1 June 1998, with a common StadtAuto/BSAG smart card as a monthly or annual pass for public transport, but it also includes the 'car-on-call', with electronically controlled access. Easy booking procedures give access to cars in the neighbourhoods 24 hours a day, 7 days a week. The basic fees are very low.

Once again, there was much scepticism to overcome. It was not easy to convince decision-makers in politics, in the government, in public transport company that car sharing was a real alternative to the private car. In fact, the European ZEUS project played a key role in opening minds to this innovative idea. Not only was the EU grant important to reduce financial risk, the fact that this was a European pilot project was important to convince key decision-makers in the public transport company. Since the initial period in 1997–98, the project has become more self-sustaining, and *Bremer Karte plus AutoCard* has won some awards.

The role of innovative public transport companies in the further development of car sharing should not be underestimated. They are a good partner—experienced with vehicle fleets, with technical maintenance, and with customer service, with public relations, and with a financial background to handle the critical phase before the break-even point. But until now most of the public transport companies have not discovered the value of integrating car sharing into intermodal mobility services.

The Bremen project *Bremer Karte plus AutoCard* gives access to the StadtAuto car-sharing vehicles by a contactless smart card (see Fig. 59). The booking process is easy: members simply phone the booking office and make a reservation for the required car at one of the StadtAuto locations. The locations are equipped with either an intelligent locker, containing the car keys, or give direct access to the

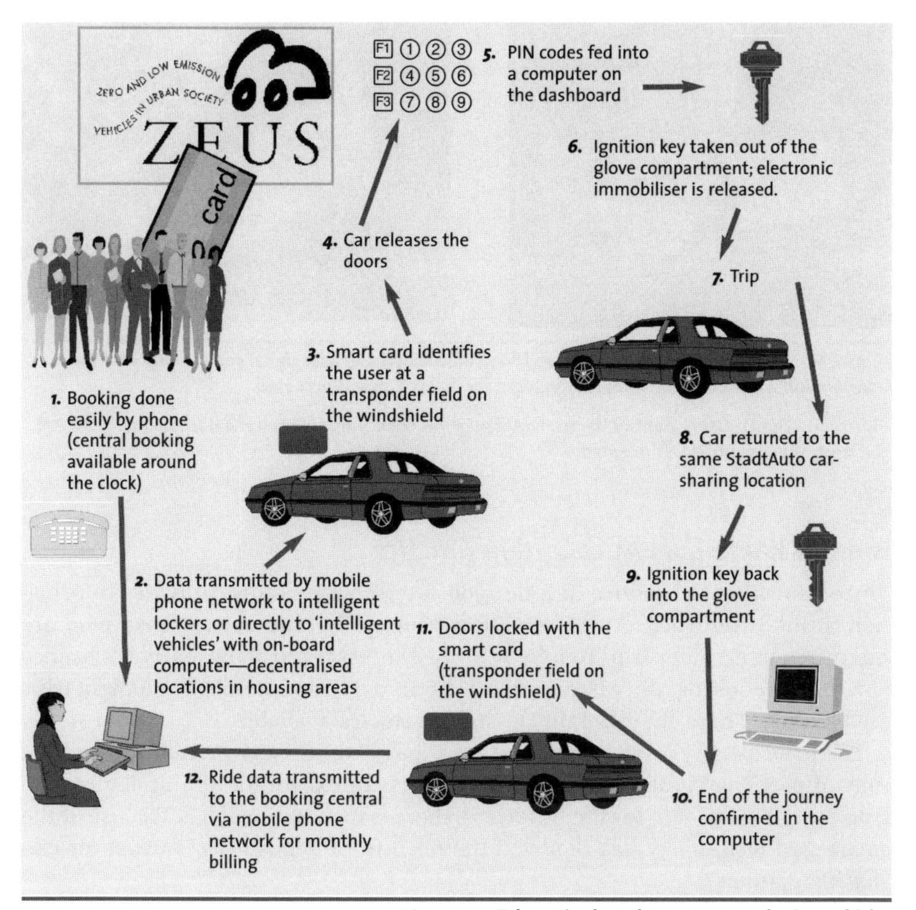

Figure 59 **Telematics-based access to car-sharing vehicles**

cars via smart card. Vehicles accessed by smart card have an on-board computer, linked via mobile phone to the central booking office, and a transponder field on the windscreen (Fig. 60). The smart card identifies the user; the computer then releases the doors, the immobiliser is released after the PIN code is entered in a small computer on the dashboard (Fig. 61). The ignition keys are in the glove compartment.

The use of innovative telematics is a prerequisite for improving the service, offering a higher level of security, easier access and an automatic billing procedure. Also, for special offers, it is necessary to have an expiry date, which is only possible with modern electronics. The electronic key for StadtAuto vehicles and the season ticket for public transport can be integrated in one card.

Access to the vehicles is easy and convenient. The short distance to the decentralised locations is a crucial element for making car sharing convenient and attractive (see Fig. 62).

Figure 60 (left) **The transponder field on the windshield: the yellow signal shows that the data transmission is working. The computer unlocks the doors if user is accepted.**

Figure 61 (right) **Users identify themselves with a PIN code (computer on the dashboard). If the PIN is correct, the immobiliser is released.**

Bremer Karte plus AutoCard *in practice*

Since launching the service in June 1998, a telephone hotline provides information around the clock. An information package is sent out to those who are interested in membership. To join, it is only necessary to show a driver's licence and ID. The booking office is available 24 hours a day, so spur-of-the-moment trips are possible. There are different classes of vehicles available.

The low-cost pay-as-you-drive system is based on the driven mileage and the time of use. The monthly basic fee is DM63 ($32). Except for an initial membership fee of DM60 ($30) for the AutoCard, there is no additional fee. The use of the smart-card technology has removed the need for a deposit (as is usual for car-sharing systems).

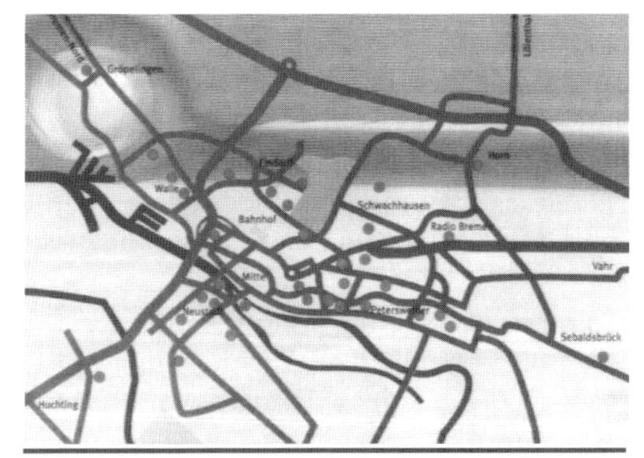

Figure 62 **In October 1999, the 46th location of StadtAuto was opened. Most stations are in the densely-populated city neighbourhoods.**

The new transport plus car-sharing offer was presented in the Werder-Bremen football stadium during half-time at a football game (December 1998).

Evaluation

The *Bremer Karte plus AutoCard* gained 500 new members in its first 12 months. A survey carried out by the University of Bremen regarding the ZEUS project showed that 150 private cars were replaced, which is the equivalent of 800 m of street space won back from the cars on the street. These 500 participants have reduced their annual car mileage by more than 800,000 km—with an equivalent CO_2 reduction of about 180 tons.

Aspects of the service which are important to new members include the following:

- No need to worry about maintenance, insurance etc. (72.3% of participants)
- Hourly rental (68.5%)
- Convenience of locations and 24 hour services (50%)
- As an alternative to the car (33.1%)

The survey not only proves the effectiveness of environmentally friendly mobility, but it also shows that a win–win situation for the public transport operator and the car-sharing organisation is possible. Sixteen per cent of clients of *Bremer Karte plus AutoCard* are also new clients for public transport season tickets. And there is also a remarkable shift from monthly season tickets (used mostly for the bad weather season only) to annual season tickets. Whereas before only 54% of new clients had annual season tickets, now that figure is 78%—giving public transport a major role as a basic means of transport.

One card fits all: modern telematics for more independence from the car

With the pilot-testing of electronic ticketing and payment systems for public transport in Bremen, the first step has been taken towards an integrated smart card with the following functions:

- Electronic purse under the CEPS (Common Electronic Purse Specifications), also readable in taxis and all electronic purse readers
- Electronic ticketing for public transport (common standard of public transport organisations and German railroad)
- Phone card
- Electronic car key for StadtAuto car-sharing vehicles
- 'Marketplaces' for integrated functions (e.g. potential miles-bonus systems, common incentives, identification for special offers, etc.)

The combined smart-card technology is part of the European INTERCEPT project—funded by the European Commission (DG XIII and DG VII).

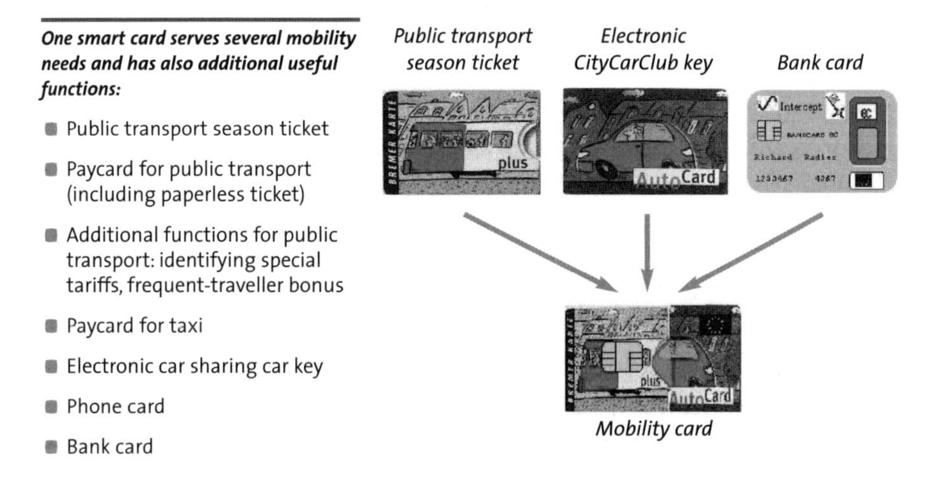

Figure 63 **Multifunctionality of the smart card**

Certification: a step to ensure high quality in car sharing

As car sharing is not legally defined, any kind of sharing a car may be called car sharing. We need some officially accepted certification. To ensure a high quality in the service and environmental quality of car sharing, in 1997 the Bremen Department for Environmental Protection launched an initiative within the German Conference of the Ministers for Environmental Protection (made up of ministers from all 16 states and the Federal government as well) to attach the eco-label, 'Blue Angel' (*blauer Umweltengel*) to car-sharing systems. This label is well known in Germany—much more so than the ISO certification, and has a strong market profile.

The Independent Jury *Umweltzeichen* (eco-label) defined the requirements for labelling car sharing.

Certification includes:

1. Service—e.g. a 24 hour service and accessibility is basic for a user-friendly service—offering a real alternative to the private car.
2. Incentives to drive less—e.g. the fee structure must not give incentive to drive more than necessary—no free mileage is allowed, a pay-as-you-drive structure should give awareness to any kilometre driven.
3. Clean vehicles—low-emission vehicles are offered.

'Blue Angel' certification will be a stimulus to the market, but certification also offers a means of solving the legal problems related to getting public space for reserved parking for car sharing.

Conclusion

Car sharing has become an integrated element of sustainable development. Future options include the replacement of parking and car ownership in new developments (housing with mobility services). Car sharing has to work under market conditions, but should get public financial support to reach critical mass, and in order to develop co-operative projects with public transport. And we still need to overcome a lot of scepticism, a problem we are facing even in cities with a well-developed and efficient transportation system.

■ Further information available at www.zeus-bremen.de

SYMBOLIC ACTS

Sometimes a symbolic act by a giant corporation can seem like a provocation, and sometimes, as in the case of Peugeot planting a small forest in a deforested area of the Amazon, the action is a significant enough symbolic act to render more credible whatever else they are doing to make their automobiles less polluting. In creating their carbon sink, Peugeot established an international partnership with French and Brazilian forestry officials, as well as with the local residents who are employed by the project. The carbon sink in the Amazon is a civic action on the part of a planetary corporate citizen. If all large corporations whose products contribute to global warming would do as much, deforestation in the Amazon or in Malaysia might be reversed. As a proviso, however, we offer an analysis of what the Peugeot carbon sink represents in terms of actual compensation for carbon emissions.

The following is a press release from Peugeot in June 1999.

PEUGEOT CREATES THE FIRST LARGE CARBON SINK

Ten million trees in the battle against global warming

Peugeot SA announced the creation of a carbon sink in the State of Mato Grosso in the heart of the Brazilian tropical rainforest. Peugeot invested 65 million francs in the project, the first of its kind, in response to the issues raised at the Kyoto and Buenos Aires conferences on global warming.

A carbon sink re-creates an ecosystem capable of absorbing large amounts of CO_2. To establish this huge environmental project, Peugeot called on the ONF (French National Bureau of Forestry), internationally known for its technical expertise and for its management of public forests, and Pro-Natura International, a Franco-Brazilian non-governmental organisation based in Paris, known for its experience in tropical rainforests and for its promotion of innovative forestry management in 25 countries.

Peugeot's carbon sink is being developed in Juruena, in the Mato Grosso, and will cover 12,000 hectares, which is twice the area of the city of Paris (about 200 km^2). It will have the capacity to absorb 50,000 tons per year of carbon, equal to 183,000 tons per year of carbon dioxide, for 40 years, at which point the full-grown trees will cease to absorb CO_2. Both internal and independent, external, auditing systems are planned for measuring actual amounts of carbon absorbed.

The sink will consist of 5,000 hectares of deforested agricultural lands, which will be replanted, and 7,000 hectares of old- and second-growth forest mixed with areas of cultivated forest. Ten million trees are being planted within the first three years of the project.

In creating this carbon sink, Peugeot is establishing the other half of their global environmental policy to complement efforts being made to reduce emissions produced by Peugeot vehicles. These efforts include cars fuelled by electricity, liquified petroleum gas, Diesel HDI (high-pressure direct-injection technology), and biofuels.

Editor's note. In Europe, an average car emits 186 g of CO_2 per km, in Japan 191 g, and in the United States 260 g. If we take 200 as the average and assume that an average car is driven 10,000 km a year and lasts for ten years, the 183,000 tons of CO_2 per year absorbed by Peugeot's carbon sink compensates for 9,000 new cars a year for 40 years, assuming no further improvements in automobile emissions occur in that period.

The 65 million francs invested by Peugeot in a carbon sink amounts to about 180 francs a car at 9,000 cars per year for 40 years—not very expensive. At that price, the logical conclusion is that all automobile manufacturers should develop carbon sinks to compensate for all the cars they sell every year worldwide, and the planet would cease warming. How big would this carbon sink have to be? If it takes 200 km^2 of new forest to compensate for 9,000 cars per year for 40 years, and Peugeot sells about 2,277,600 (1998) new cars and light-utility vehicles a year worldwide, full compensation for emissions would require a carbon sink of approximately 40,000 km^2. The world's automakers sold more than 51 million new cars and light trucks worldwide in 1998. To compensate for their emissions, assuming no future reductions, it would take a carbon sink of new forest equal to 15% of Brazil to get us through the next 40 years. And this is without addressing emissions from the more than 500 million cars already on the roads of the world, or, it is important to point out, without taking note of the size of the world's already-existing and still-active carbon sinks or of new ones being planted.

Obviously, and Peugeot understands this very well, while their Mato Grosso carbon sink is a grand gesture that other automobile manufacturers might want to join, most of the challenge lies in reducing fuel consumption, in changing the ways cars can be driven, and in reducing car use altogether.

Regardless of its imprecise value, the carbon sink idea is spreading as it becomes a tool of the carbon credit process now developing. Tokyo Electric Power Co. is planting 40,000 hectares of new forest in Australia to reduce the impact of its carbon emissions. Tokyo Electric's carbon credit deal signed with State Forests of New South Wales is worth US$81 million.

BIBLIOGRAPHY

Adriaanse, A., S. Bringezu, A. Hammond, Y. Moriguchi, E. Rodenburg, D. Rogich and H. Schütz (1997) *Resource Flows: The Material Basis of Industrial Economies* (Washington, DC: World Resources Institute; Basel: Wuppertal Institute/Birkhäuser Verlag; The Hague: Netherlands Ministry of Housing, Spatial Planning and Environment; Japan: National Institute for Environmental Studies).

Adriaanse, A., S. Bringezu, A. Hammond, Y. Moriguchi, E. Rodenburg, D. Rogich and H. Schütz (1998) *Stoffströme: Die materielle Basis von Industriegesellschaften* (German rev. edn; Washington, DC: World Resources Institute; Basel: Wuppertal Institute/ Birkhäuser Verlag; The Hague: Netherlands Ministry of Housing, Spatial Planning and Environment; Japan: National Institute for Environmental Studies).

Aggeri, F., and A. Hatchuel (1998) 'Managing Creation and Learning of New Expertise in Automobile Development Projects', in R.A. Lundin and C. Midler (eds.), *Projects as Arenas for Renewal and Learning Processes* (Dordrecht, Netherlands: Kluwer Academic Publishers).

Air Conditioning, Heating and Refrigeration News (1993) 'Greenpeace and Refrigerants: Can advocacy go too far?', *Air Conditioning, Heating and Refrigeration News*, 13 December 1993: 16.

Anastas, P.T., and J.J. Breen (1997) 'Design for the Environment and Green Chemistry: The Heart and Soul of Industrial Ecology', *Journal of Cleaner Production* 5.1–2.

Anderson, R.C. (1998) *Mid-Course Correction* (White River Junction, VT: Chelsea Green Publishing).

Arora, S., and T.N. Carson (1995) 'An Experiment in Voluntary Environmental Regulation: Participation in EPA's 33/50 Program', *Journal of Environmental Economics and Management* 28 (May 1995): 271-86.

Aspen Institute (1999) *The Alternative Path: A Cleaner, Cheaper Way to Protect and Enhance the Environment* (Series on the Environment in the 21st Century; Washington, DC: The Aspen Institute).

Ayres, R.U., W.H. Schlesinger and R.H. Socolow (1994) 'Human Impacts on the Carbon and Nitrogen Cycles', in R. Socolow, C. Andrews, F. Berkhout and V. Thomas (eds.), *Industrial Ecology and Global Change* (Cambridge, UK: Cambridge University Press).

Baskin, Y. (1997) *The Work of Nature: How the Diversity of Life Sustains Us* (Washington, DC: Island Press).

Beste, D. (1994) 'The Greenfreeze Campaign', *Akzente*, December 1994: 26-29.

Böge, S. (1993) *Road Transport of Goods and the Effects on the Spatial Environment* (Wuppertal, Germany: Wuppertal Institute, July 1993).

Boone, L.E., and D.L. Kurtz (1998) *Contemporary Marketing Wired* (Fort Worth, TX: Dryden Press, 9th edn).

Bourne, G. (1969) *Change in the Village* (New York: Augustus M. Kelley).

Bringezu, S. (1997) 'Accounting for the Physical Basis of National Economies: Material Flow Indicators', in B. Moldan, S. Billharz and R. Matravers (eds.), *SCOPE 58: Sustainability Indicators* (Paris: Scientific Committee on Problems of the Environment [SCOPE]): 170-80B.

Brown, L.D. (1991) 'Bridging Organizations and Sustainable Development', *Human Relations* 44.8: 807-31.

Busck, O., and S. Handberg (1995) 'Employee Participation as a Means to Achieve Organically Integrated Dynamics and Continued Results in Companies Aiming at Cleaner Production', *Proceedings from the Second International Conference on Waste Minimization and Clean Technologies*, Barcelona, 1995. *Miljø og beskæftigelse: Erhvervsmæssige perspektiver* (ed. O. Busck; Copenhagen: Specialarbejderforbundet i Danmark [SiD]).

Business and the Environment (1994) 'Environmental group sets strategies', *Business and the Environment* 12.5 (December 1994).

Butz, C., and A. Plattner (1999) *Sustainable Investments: An Analysis of Returns in Relation to Environmental and Social Criteria* (Sarasin Basic Report; Basel: Bank Sarasin & Co.).

Capra, F., and G. Pauli (eds.) (1995) *Steering Business toward Sustainability* (Tokyo: United Nations University Press).

Cerne, O., and A.B. Antonsson (1999) *Vad betyder miljöledningssystem för arbetsmiljön?* (IVL-Rapport B1341; Stockholm: IVL Svenska Miljöinstitutet).

Choong, H. (1996) 'Procurement of Environmentally Responsible Material Program: Eco-efficient Concepts for the Electronics Industry towards Sustainability', *Proceedings of Care Vision 2000, Care Innovation '96*, Frankfurt a.M. 18–20 November 1996: 258-62.

Christensen, P., *et al.* (1996) 'Implementing EMS in Danish Industry', *Eco-Management and Auditing* 3: 56-62.

Colborn, T., D. Dumanoski and J. Peterson Myers (1996) *Our Stolen Future: Are we threatening our fertility, intelligence and survival? A Scientific Detective Story* (East Rutherford, NJ: Dutton Publishing).

Commission of the European Communities (1996) *Communication from the Commission to the Council and the European Parliament on Environmental Agreements*, COM(96) 561 final.

Corder, M. (1997) 'Greenpeace no longer world's savior group', *Herald Journal* (Logan, UT), 7 October 1997: 17.

Dalton, A.J.P. (1998) *Safety, Health and Environmental Hazards at the Workplace* (London: Cassell).

Daly, H.E., and J. Cobb (1989) *For the Common Good: Redirecting the Economy towards Community, the Environment and Sustainable Development* (Boston, MA: Beacon Press).

Daly, H.E. (1996) *Beyond Growth: The Economics of Sustainable Development* (Boston, MA: Beacon Press).

Davies, J.C., and J. Mazurek (1998) *Pollution Control in the United States: Evaluating the System* (Washington, DC: Resources for the Future).

Dechant, K., and B. Altman (1994) 'Environmental Leadership: From Compliance to Competitive Advantage', *Academy of Management Executive* 8: 7-20.

Department of Social Security, UK (1998) *A New Contract for Welfare: Partnership in Pensions* (London: Department of Social Security, December 1998).

de Romana, A.L. (1989) *The Autonomous Economy: An Emerging Alternative to Industrial Society* (*INTERculture* [International Journal of Intercultural and Transdisciplinary Research]; Montreal: Monchanin Cross-Cultural Centre).

DETR (UK Department of the Environment, Transport and the Regions) (1998) *Sustainability Counts* (consultation paper on a set of 'headline' indicators of sustainable development; London: DETR).

Die Welt (1996) 'Germany: Foron Hausgeräte files for bankruptcy', *Die Welt*, 16 March 1996: 15.

Douthwaite, R. (1992) *The Growth Illusion: How Economic Growth Enriched the Few, Impoverished the Many and Endangered the Planet* (Totnes, UK: Green Books).

Douthwaite, R. (1996) *Short Circuit: Strengthening Local Economies for Security in an Unstable World* (Totnes, UK: Green Books).

Douthwaite, R. (1999) *The Ecology of Money* (Totnes, UK: Green Books).

Dowie, M. (1995) *Losing Ground: American Environmentalism at the Close of the Twentieth Century* (Cambridge, MA: MIT Press).

EEA (European Environment Agency) (1998a) *Technical Report on the Fisheries Situation in the EU* (prepared for the EEA by J. Pope; Copenhagen: EEA).

EEA (European Environment Agency) (1998b) *Background Paper on Eco-Efficiency* (D. Gee and S. Moll; background paper for EEA workshop 'Making Sustainability Accountable: Eco-Efficiency, Resource Productivity and Innovation', Copenhagen, 28–30 October 1998; Copenhagen: EEA).

EEA (European Environment Agency) (1998c) *Life Cycle Assessment (LCA): A Guide to Approaches, Experiences and Information Sources* (Environmental Issues Series no. 6; Copenhagen: EEA; Luxembourg: Office for Official Publications of the European Communities).

EEA (European Environment Agency) (1999a) *Environment in the European Union at the Turn of the Century* (Copenhagen: EEA, June 1999).

EEA (1999b) *The Typology of Indicators* (Copenhagen: EEA).

EEA (1999c) *Progress with Integration: A Contribution to the Global Assessment of the Fifth Environmental Action Programme* (Copenhagen: EEA).

EEA (European Environment Agency)/UNEP (United Nations Environment Programme) (1998) *Chemicals in Europe: Low Doses, High Stakes* (Annual Message 2 on the State of Europe's Environment; Copenhagen: EEA; Geneva: UNEP).

Enterprise for the Environment (1998) *The Environmental Protection System in Transition: Toward a More Desirable Future* (Washington, DC: Center for Strategic and International Studies).

EPA (US Environmental Protection Agency) (1995) *An Introduction to Environmental Accounting as a Business Tool Management: Key Concepts and Terms* (EPA 742-R-95-001; Washington, DC: EPA Office of Pollution Prevention and Toxics).

EPA (US Environmental Protection Agency) (1996) *Environmental Cost Accounting Case Studies: Full Cost Accounting for Decision Making at Ontario Hydro* (EPA 742-R-95-004; Washington, DC: EPA Office of Pollution Prevention and Toxics).

EPA (US Environmental Protection Agency) (1998) *Partners for the Environment: Collective Statement of Success* (Washington, DC: EPA Office of Reinvention).

Erkman, S. (1995) *Ecologie industrielle, métabolisme industriel, et société d'utilisation* (Paris: Foundation for the Progress of Humanity).

Erkman, S. (1997) 'Industrial Ecology: A Historical View', *Journal of Cleaner Production* 4.4.

European Commission (1993) *White Paper on Growth, Competitiveness, and Employment* (COM[93]700 final; Brussels: European Commission, 5 December 1993).

European Commission (1997) *Functional Implications of Biodiversity in Soils* (Ecosystems Research Report, 24; Brussels: European Commission).

European Commission (1998a) *Understanding Biodiversity* (Ecosystems Research Report, 25; Brussels: European Commission).

European Commission (1998b) *Towards Fair and Efficient Pricing of Transport* (Brussels: European Commission).

European Commission, DG XI (1999) *Workshop on Integrated Product Policy*, 8 December 1998, Final Report.

Federal Ministry for the Environment, Germany (1998) *Draft Programme for Priority Areas in Environmental Policy* (Bonn: Federal Ministry for the Environment).

Feldman, I. (1997) 'ISO 14000 can underpin a new "dual-track" regulatory system: Greentracking as an Alternative to Command and Control', *Environmental Business Journal* 11–15.

Fischer, K., and J. Schot (eds.) (1993) *Environmental Strategies for Industry* (Washington, DC: Island Press).

Fussler, C., with P. James (1996) *Driving Eco-Innovation: A Breakthrough Discipline for Innovation and Sustainability* (London: Pitman).

Gallarotti, F.M. (1995) 'It pays to be green: The Managerial Incentive Structure and Environmentally Sound Strategies', *Columbia Journal of World Business* 30: 38-57.

Gee, D. (1994) *Clean Production* (London: Manufacturing, Science and Finance Union).

Gitlin, T. (1995) *The Twilight of Common Dreams: Why America is Wracked by Culture Wars* (New York: Metropolitan Books).

Gitlin, T. (1999) *Sacrifice: A Novel* (New York: Metropolitan Books).

Gitlin, T. (2001) *Media Unlimited: Life in the Torrent of Image and Sound* (New York: Metropolitan Books/Henry Holt).
Goering, P., H. Norberg Hodge and J. Page (1993) *From the Ground Up: Rethinking Industrial Agriculture* (London: Zed Books).
Goodland, R. (1991) 'The Case that the World Has Reached its Limits', in *Environmentally Sustainable Economic Development: Building on Brundtland* (Paris: United Nations Educational, Scientific and Cultural Organisation).
Grieder, W. (2000) 'Business Creates Eco-Side', *The Nation*, 28 February 2000.
Hamilton, K., G. Atkinson, D. Pearce and C. Serge (1997) *Genuine Savings as an Indicator of Sustainability* (Working Paper GEC97-03; Norwich, UK: School of Environmental Sciences, University of East Anglia).
Handberg, S. (1993) *Arbejdsmiljø og renere teknologi i fiskeindustrien, Arbejdsmiljøfondet* (Copenhagen).
Handelsblatt (1995) 'Germany: Samsung withdraws from Foron deal', *Handelsblatt*, 10 March 1995: 15.
Handelsblatt (1996) 'Germany: Foron in shock after losing Turkish buyer', *Handelsblatt*, 21 February 1996: 12.
Hartman, C.L., and E.R. Stafford (1997) 'Market-Based Environmentalism: Developing Green Marketing Strategies and Relationships', in D.T. Leclair and M. Hartline (eds.), *American Marketing Association Winter Educators' Conference Proceedings* (Chicago: American Marketing Association): 156-63.
Hartman, C.L., E.R. Stafford and M.J. Polonsky (1999) 'Green Alliances: Environmental Groups as Strategic Bridges to Other Stakeholders', in M. Charter and M.J. Polonsky (eds.), *Greener Marketing: A Global Perspective on Greening Marketing Practice* (Sheffield, UK: Greenleaf Publishing): 164-80.
Hausker, K. (1999a) 'Reinventing Environmental Regulation: The Only Path to a Sustainable Future', *Environmental Law Reporter.*
Hausker, K. (1999b) *The Convergence of Ideas on Improving the Environmental Protection System* (Washington, DC: Center for Strategic and International Studies).
Hawken, P., A. Lovins and L.H. Lovins (1999) *Natural Capitalism: Creating the Next Industrial Revolution* (New York: Little Brown & Co.).
Hemphill, T. (1996) 'The New Era of Business Regulation', *Business Horizons* 39 (July/August 1996): 26-30.
Hersh, R. (1996) *A Review of Integrated Pollution Control Efforts in Selected Countries* (Discussion Paper 97-15; Washington, DC: Resources for the Future).
Hicks, J.R. (1939) *Value and Capital* (Oxford, UK: Oxford University Press [2nd edn 1946]).
Houghton, J.T. (1994) *Global Warming: The Complete Briefing* (Elgin, IL/Oxford, UK: Lion Publishing [repr. Cambridge, MA: Cambridge University Press, 1997]).
Hueting, R. (1980) *New Scarcity and Economic Growth* (New York: Oxford University Press).
IEA (International Energy Agency) (1998) *World Energy Outlook* (Paris: IEA, November 1998).
Irish Times (1999) 'Wal-Mart buy may begin wave of European supermarket mergers', *The Irish Times*, 2 July 1999.
ISO (International Organisation for Standardisation) (1997) *ISO 14040: Environmental Management, Life Cycle Assessment, Principles and Framework* (Geneva: ISO).
ISO (International Organisation for Standardisation) (1998) *ISO DIS 14021: Label Type III, Environmental Labels and Declarations, Self-Declared Environmental Claims* (Geneva: ISO).
Jackson, T. (1996) *Material Concerns: Pollution, Profit and Quality of Life* (London: Routledge).
Jackson, T., N. Marks, J. Ralls and S. Stymne (1997) *Sustainable Economic Welfare in the UK 1950–1996* (London: New Economics Foundation).
Juutinen, A., and I. Mäenpää (1999) *Time Series for the Total Material Requirement of Finnish Economy: Summary* (Eco-efficient Finland Project, interim report, 15 August 1999; University of Oulu, Thule Institute, www.thule.oulu.fi/ecoef).
Kalke, M. (1994) 'The Foron-Story', *Akzente*, December 1994: 20-25.

Kamp, A. (1997) *Integreret miljø- og arbejdsmiljøarbejde på danske virksomheder* (Lyngby, Denmark: Institut for teknologi og samfund, Danmarks Tekniske Universitet).

Kofoed, N.M., Petersen and U. Sæbye (1998) *Medarbejdernes frie projektarbejder på virksomheden* (Roskilde: Roskilde University).

Kox, H.L.M., and H. Linnemann (1994) 'International Commodity-Related Agreements as Instruments for Promoting Sustainable Production of Primary Export Commodities', in *Sustainable Resources Management and Resources Use: Policy Questions and Research Needs* (Rijswijk, Netherlands: RMNO).

Lazarus, O.S. (1991) 'Save our Soils', *Our Planet* 2.4.

Levene, A. (1994) 'Greenpeace uses the market to make business "green up" its act', Reuters World Service, 13 March 1994.

Levy, D.L. (1997) 'Business and International Environmental Treaties: Ozone Depletion and Climate Change', *California Management Review* 39.2: 54-71.

Liedtke, C., T. Orbach and H. Rohn (1994) 'Towards a Sustainable Company: Resource Management at the Kambium Furniture Workshop', in *Analysis for Action: Support for Policy towards Sustainability by Material Flow Accounting* (Proceedings of the Con-account Conference; Wuppertal, Germany: Wuppertal Institute).

LO (Landsorganisationen Danmark [Danish National Trade Union Federation]) (1998) *Employee Participation: A Resource in Environmental Management* (Copenhagen: LO).

Lober, D.J. (1997) 'Explaining the Formation of Business–Environmentalist Collaborations: Collaborative Windows and the Paper Task Force', *Policy Sciences* 30.1: 1-24.

Long, F.J., and M.B. Arnold (1995) *The Power of Environmental Partnerships* (Fort Worth, TX: Dryden Press).

Lorentzen, B., A. Remmen, L. Nielsen and P.T. Aldrich (1997) *Medarbejderdeltagelse ved indførelse af renere teknologi: Hovedrapport* (Miljøprojekt no. 354; Copenhagen: Miljøstyrelsen [Danish Environmental Protection Agency]).

Louppe, A., and A. Rocaboy (1994) 'Green Consumerism and Marketing', *French Management Review* 98 (March–May 1994).

Lovejoy, T.E. (1995) 'Will Expectedly the Top Blow Off?', *Bioscience, Science and Biodiversity Policy Supplement* S-3-6.

Lowe, E.A. (1997) 'Creating By-product Resource Exchanges: Strategies for Eco-industrial Parks', *Journal of Cleaner Production* 5.1–2.

Marshall, A. (1920) *Principle of Economics* (London: Macmillan).

Mazurek, J. (1999a) *Making Microchips: Policy, Globalisation, and Economic Restructuring in the Semiconductor Industry* (Cambridge, MA: MIT Press).

Mazurek, J. (1999b) *The Use of Voluntary Agreements in the United States: An Initial Survey* (ENV/EPOC/GEEI[98]27/FINAL; Paris: Organisation for Economic Co-operation and Development).

McCloskey, M. (1992) 'Twenty Years of Change in the Environmental Movement: An Insider's View', in R.E. Dunlap and A.G. Mertig (eds.), *American Environmentalism: The US Environmental Movement, 1970–1990* (Philadelphia: Taylor & Francis): 77-88.

Meadows, D.H., D.L. Meadows, J. Randers and W.W. Behrens (1972) *The Limits to Growth: A Report for the Club of Rome's Project on the Predicament of Mankind* (London: Pan).

Mendleson, N., and M.J. Polonsky (1995) 'Using Strategic Alliances to Develop Credible Green Marketing', *Journal of Consumer Marketing* 12.2: 4-18.

Menon, A., and A. Menon (1997) 'Enviropreneurial Marketing Strategy: The Emergence of Corporate Environmentalism as Market Strategy', *Journal of Marketing* 61.1: 51-67.

Metzenbaum, S. (1998) *Making Measurement Matter: The Challenge and Promise of Building a Performance-Focused Environmental Protection System* (Washington, DC: Brookings Institution Center for Public Management).

Mill, J.S. (1848) *Principles of Political Economy, with Some of their Applications to Social Philosophy* (London: John W. Parker).

Milne, G.R., E.S. Iyer and S. Gooding-Williams (1996) 'Environmental Organization Alliance Relationships within and across Nonprofit, Business, and Government Sectors, *Journal of Public Policy and Marketing* 15.2 (Fall 1996): 203-15.

Ministry of the Environment, Sweden (1998) *Key Indicators for Ecologically Sustainable Development* (proposal from the Swedish Environmental Advisory Council; Stockholm: Ministry of the Environment).

Mündl, A., *et al.* (1999) *Sustainable Development by Dematerialization in Production and Consumption: Strategy for the New Environmental Policy in Poland* (Report 3; Warsaw: Institute for Sustainable Development).
Murphy, D.F., and J. Bendell, (1997) *In the Company of Partners: Business, Environmental Groups, and Sustainable Development Post-Rio* (Bristol, UK: The Policy Press).
National Association of Attorneys-General (1990) *The Green Report: Findings and Preliminary Recommendations for Responsible Advertising* (Washington, DC: NAAG Publications, December 1990).
Nixon, W. (1993) 'A Breakfast among Peers: Environmental whistleblowers have some stories to tell', *E Magazine*, September/October 1993: 14-19.
NRC (National Research Centre) (1992) *Restoration of Aquatic Ecosystems: Science, Technology and Public Policy* (Washington, DC: National Research Centre, National Academy Press).
OECD (Organisation for Economic Co-operation and Development) (1998a) *Eco-efficiency* (Paris: OECD).
OECD (Organisation for Economic Co-operation and Development) (1998b) *Voluntary Approaches for Environmental Protection in OECD Countries* (ENV/EPOC/GEEI[98]29/FINAL; Paris: OECD).
Orton, D. (1994) *Struggling against Sustainable Development: A Canadian Perspective* (Boston: Z Papers).
ÖTV-Union (1995) *Rationelle Energieausnutzung in öffentliche Gebäuden* (Stuttgart: ÖTV-Union).
Pauli, G. (1997) 'Zero Emissions: The Ultimate Goal of Cleaner Production', *Journal of Cleaner Production* 5.1–2.
PCSD (President's Council on Sustainable Development) (1996) *Sustainable America: A New Consensus* (Washington, DC: PCSD).
PCSD (President's Council on Sustainable Development) (1999) *Towards a Sustainable America: Advancing Prosperity, Opportunity, and a Healthy Environment for the 21st Century* (Washington, DC: Government Printing Office).
Polonsky, M.J. (1996) 'Stakeholder Management and the Stakeholder Matrix: Potential Strategic Marketing Tools', *Journal of Market Focused Management* 1.3: 209-29.
Polonsky, M.J., H. Suchard and D. Scott (1997) 'A Stakeholder Approach to Interacting with the External Environment', in P. Reed, S. Luxton and M. Shaw (eds.), *Proceedings of the 1997 Australian and New Zealand Marketing Educators' Conference* (Melbourne: ANZMA): 495-508.
Porter, M.E., and C. van der Linde (1995) 'Green and Competitive: Ending the Stalemate', *Harvard Business Review* 73.5 (September/October 1995): 120-33.
Reitman, V. (1997) 'Ford is investing in Daimler-Ballard fuel-cell venture', *The Wall Street Journal*, 16 December 1997: B4.
RMNO (Advisory Council for Research on Nature and the Environment) (1994) *Sustainable Resource Management and Resource Use: Policy Questions and Research Needs* (Rijswijk, Netherlands: RMNO).
Roberts, J.A. (1996) 'Green Consumers in the 1990s: Profile and Implications for Advertising', *Journal of Business Research* 26.2: 217-31.
Rohrschneider, R. (1991) 'Public Opinion toward Environmental Groups in Western Europe: One Movement or Two?', *Social Science Quarterly* 72 (June 1991): 251-66.
Rokeach, M. (1973) *The Nature of Human Values* (New York: Free Press).
Romm, J. (1999) *Cool Companies: How the Best Businesses Boost Profits and Productivity by Cutting Greenhouse Gas Emissions* (Washington, DC: Island Press).
Rowley, T. (1997) 'Moving Beyond Dyadic Ties: A Network Theory of Stakeholders' Influence', *Academy of Management Review* 22.4: 887-910.
RS (Royal Society)/NAS (US National Academy of Sciences) (1992) *Population Growth, Resource Consumption and a Sustainable World* (London: RS; Washington: NAS).
Russo, M.V., and P.A. Fouts (1997) 'A Resource-Based Perspective on Corporate Environmental Performance and Profitability', *Academy of Management Journal* 40.3: 534-59.
Sarkin, D., and J. Schelkin (1991) 'Environmental Concerns and the Business of Banking', *The Journal of Commercial Bank Lending* 73.3: 7-19.

Schmidheiny, S., with the Business Council for Sustainable Development (ed.) (1992) *Changing Course: A Global Business Perspective on Development and the Environment* (Cambridge, MA: MIT Press).

Schumacher, E.F. (1973) *Small is beautiful: Economics as if People Mattered* (London: Blond & Briggs).

Schuster, M. (1997) 'Translating Material Flow Analysis into Environmental Policy in Austria', in *Analysis for Action: Support for Policy towards Sustainability by Material Flow Accounting* (Proceedings of the Conaccount Conference; Wuppertal, Germany: Wuppertal Institute).

Schwarz, E.J., and K.W. Steininger (1997) 'Implementing Nature's Lesson: The Industrial Recycling Network Enhancing Regional Development', *Journal of Cleaner Production* 5.1–2.

Sen, A. (1999) *Development as Freedom* (New York: Knopf)

Sharma, S., H. Vredenburg and F. Westley (1994) 'Strategic Bridging: A Role for the Multinational Corporation in Third World Development', *Journal of Applied Behavioral Science* 30.4: 458-76.

Speth, J.G. (1989) 'A Luddite Recants: Technological Innovation and the Environment', in *The Amicus Journal,* Spring 1989.

Stafford, E.R., and C.L. Hartman (1996) 'Green Alliances: Strategic Relations between Businesses and Environmental Groups', *Business Horizons* 39.2: 50-59.

Sterner, T. (ed.) (1999) *The Market and the Environment: The Effectiveness of Market-Based Policy Instruments for Environmental Reform* (Cheltenham, UK: Edward Elgar).

Stewart, R.B. (1993) 'Economic Competitiveness and the Law', *Yale Law Review* 102 (June 1993): 2039-2106.

SustainAbility/UNEP (United Nations Environment Programme) (1998) *The Non-Reporting Report* ('Engaging Stakeholders' series; London: SustainAbility/UNEP).

Tundo, P., and J.J. Breen (1999) 'Venice: A Center for Green Chemistry on the Continent', *Today's Chemist at Work* 8.2: 52-59

UNDP (United Nations Development Programme) (1992) *Human Development Report 1992* (New York: Consumption for Human Development).

UNDP (United Nations Development Programme) (1998) *Human Development Report 1998* (New York: Consumption for Human Development, www.undp.org/hdro/98.htm).

Vidal, J. (1992) 'The Big Chill', *The Guardian,* 19 November 1992: 2.

von Weizsäcker, E.U., A.B. Lovins and L.H. Lovins (1997) *Factor Four: Doubling Wealth, Halving Resource Use* (The New Report to the Club of Rome; London: Earthscan Publications).

VROM (Ministerie van Volkshiusvesting, Ruimtelijke Ordening en Miieubeheer [Ministry of Housing, Spatial Planning and the Environment, Netherlands]) (1998) *National Environmental Policy Plan 3* (The Hague: VROM).

Walsh, M.W. (1995) 'Case Study: Company puts faith in freon-free fridge', *Los Angeles Times,* 10 January 1995: 4.

Wasik, J.F. (1996) *Green Marketing and Management: A Global Perspective* (Cambridge, MA: Blackwell).

WCED (World Commission on Environment and Development) (1987) *Our Common Future* ('The Brundtland Report'; Oxford, UK: Oxford University Press).

Westley, F., and H. Vredenburg (1991) 'Strategic Bridging: The Collaboration between Environmentalists and Business in the Marketing of Green Products', *Journal of Applied Behavioral Science* 27.1: 65-90.

Zwetsloot, G., and J. Bos (1998) *Design for Sustainable Development: Environmental Management and Safety and Health* (Dublin: European Foundation).

ABBREVIATIONS

5EAP	5th Environmental Action Programme
ABB	Asea Brown Boveri
ADEME	Agence de l'Environnement et de la Maîtrise de l'Energie (France)
AFL	American Federation of Labor
AOX	adsorbable organic halogen
ASEAN	Association of South-East Asian Nations
BAT	best available technology
BATNEEC	best available technology not entailing excessive costs
BDO	1,4-butanediol
BSCD	Business Council for Sustainable Development
CAFE	Corporate Average Fuel Economy
CAGR	compound annual growth rate
CDM	clean development mechanism
CEO	chief executive officer
CEPS	Common Electronic Purse Specifications
CERES	Coalition of Environmentally Responsibility Economies
CFC	chlorofluorocarbon
CH_3COOH	acetic acid
CIO	Congress of Industrial Organisations
CIS	Commonwealth of Independent States
CO	carbon monoxide
CO_2	carbon dioxide
CSD	Commission on Sustainable Development (United Nations)
CVT	continuously variable transmission
CZ	Campana-Zarate
DDT	dichlorodiphenyltrichloroethane
DfD	design for dismantling
DfE	design for environment
DfR	design for recycling
DM	deutsche mark
DMI	direct material input
DoE	Department of Energy (USA)
DPSIR	Driving forces–Pressures–State–Impact–Responses
EB	electron beam
EC	European Commission
ECMT	European Conference of Ministers of Transport
ECP	environmentally conscious product
EDF	Electricité de France

EEA	European Environment Agency
EEC	European Economic Community
EH&S	environmental health and safety
EIC	Economic Instruments Collaborative (Canada)
EIME	Environmental Information and Management Explorer
ELP	Environmental Leadership Program of the US EPA
EMAS	Eco-management and Audit Scheme
EMS	environmental management system
EPA	US Environmental Protection Agency
EPE	European Partners for the Environment
ESCO	energy service company
EU	European Union
EV	electric vehicle
GATT	General Agreement on Tariffs and Trade
GBL	gamma butyrolactone
GDP	gross domestic product
GEF	Global Environment Facility
GHG	greenhouse gas
GM	General Motors
GNP	gross national product
GRI	Global Reporting Initiative
H_2SO_4	sulphuric acid
HAP	hazardous air pollutant
HC	hydrocarbon
HCl	hydrogen chloride (in solution, hydrochloric acid)
HDI	high-pressure direct-injection technology
HEI	Hyundai Electronics Industries Co. Ltd
HNO_3	nitric acid
ICLEI	International Environmental Agency for Local Governments
IEA	International Energy Agency
IFC	International Finance Corporation
IM	information management
IMF	International Monetary Fund
INESTENE	Institut d'Evaluation des Stratégies sur l'Energie et l'Environnement (France)
IPA	isopropyl alcohol
IPCC	Intergovernmental Panel on Climate Change
ISO	International Organisation for Standardisation
JAMA	Japan Automotive Machinery and Tool Manufacturers' Association
JIS	Japanese International Standard
LA 21	Local Agenda 21
LCA	life-cycle assessment
LCI	Lean Construction Institute
LCP	least-cost planning
LETS	Local Exchange Trading System
LO	Landsorganisationen Danmark
MITI	Ministry of International Trade and Industry
MOU	memorandum of understanding
N_2O	nitrous oxide
NACEPT	National Advisory Council for Environmental Policy and Technology (USA)
NAFTA	North American Free Trade Agreement
NaOH	sodium hydroxide
Neq	noise equivalent
NGO	non-governmental organisation
NH_3	ammonia

NIMBY 'not in my back yard'
NO_x nitrogen oxides
OD ozone depletion
ODA overseas development assistance
ODS ozone-depleting substance
OECD Organisation for Economic Co-operation and Development
OEM original equipment manufacturer
ONF Office National des Forêts (France)
OPEC Organisation of Petroleum Exporting Countries
P&EM Photopolymer and Electronic Materials
P2 pollution prevention
PCB polychlorinated biphenyl
PCSD President's Council on Sustainable Development
PEM proton exchange membrane
PERM Procurement of Environmentally Responsible Material
PET polyethylene terephthalate
PIN personal identification number
PM particulate matter
POP persistent organic pollutant
PRTR Pollutant Release and Transfer Register
PVC polyvinyl chloride
QMS quality management system
R&D research and development
RDF refuse-derived fuel
RSRC Regulated Substance and Recyclability Certification
S&P Standard & Poor's Corp.
SA succinic acid
SEDIF Syndical d'Eau de l'Ile de France
SEP superior environmental performance
SiD Specialarbejderforbundet i Danmark
SME small and medium-sized enterprise
SMI small and medium-sized industry
SO_2 sulphur dioxide
SO_x oxides of sulphur
SVA shareholder value added
TAG Technical Advisory Group
THF tetrahydrofuran
TMR total material requirement
TRI Toxics Release Inventory
UNCTAD United Nations Conference on Trade and Development
UNDP United Nations Development Programme
UNEP United Nations Environment Programme
UNFCCC United Nations Framework Convention on Climate Change
UNIDO United Nations Industrial Development Organisation
USCAR United States Council for Automotive Research
VOC volatile organic compound
VP vice president
VVT-i variable valve timing-intelligent
WBCSD World Business Council for Sustainable Development
WCED World Commission on Environment and Development
WTO Word Trade Organisation
WWF World Wide Fund for Nature
XL eXcellence and Leadership project of the US EPA
ZAC zone d'aménagement concerte
ZI zone industrielle